THE MATHEMATICS OF COMBUSTION

JOHN D. BUCKMASTER, EDITOR

PHILADELPHIA
1985

Copyright © 1985 by Society for Industrial and Applied Mathematics.
All rights reserved.

Library of Congress Catalog Card Number: 85-50339
ISBN: 0-89871-053-7

CONTENTS

Contributors vii

Foreword .... *H. T. Banks*   ix

Preface .... *John D. Buckmaster*   xi

Chapter I.   An Introduction to Combustion Theory
    .... *John D. Buckmaster*   3

Chapter II.   Sensitivity Analysis of Combustion Systems
    .... *Herschel Rabitz*   47

Chapter III.   Turbulent Combustion .... *F. A. Williams*   97

Chapter IV.   Detonation in Miniature .... *Wildon Fickett*   133

Chapter V.   Finite Amplitude Waves in Combustible Gases
    .... *J. F. Clarke*   183

Index 247

CONTRIBUTORS

JOHN D. BUCKMASTER is Professor of Aeronautical and Astronautical Engineering, University of Illinois, Urbana, Illinois.

J. F. CLARKE is Professor of Theoretical Gas-Dynamics, College of Aeronautics, Cranfield Institute of Technology, Cranfield, Bedford, England.

WILDON FICKETT is with Los Alamos National Laboratory, Los Alamos, New Mexico.

HERSCHEL RABITZ is Professor of Chemistry, Princeton University, Princeton, New Jersey.

F. A. WILLIAMS is Professor of Mechanical and Aerospace Engineering, Princeton University, Princeton, New Jersey.

FOREWORD

This is the second volume in the SIAM series *Frontiers in Applied Mathematics*. This continuing series will focus on "hot topics" in applied mathematics, and will consist of, in general, unrelated volumes, each dealing with a particular research topic that should be of significant interest to a spectrum of members of the scientific community. Distinguished scientists and applied mathematicians will be solicited to contribute their points of view on "state-of-the-art" developments in the topics addressed.

The volumes are intended to provide provocative intellectual forums on emerging or rapidly developing fields of research as well as be of value in the general education of the scientific community on current topics. In view of this latter goal, the solicited articles will be designed to give the nonexpert, nonspecialist some appreciation of the goals, problems, difficulties, possible approaches and tools, and controversial aspects, if any, of current efforts in an area of importance to scientists of varied persuasions.

Each volume will begin with a tutorial article in which technical terms, jargon, etc. are introduced and explained. This will be followed by a number of research-oriented summary contributions on topics relevant to the subject of the volume. We hope that the presentations will give mathematicians and nonmathematicians alike some understanding of the important role mathematics is playing, or perhaps might play, in what academicians often euphemistically call "the real world." We therefore expect each volume to contribute some further understanding of the important scientific interfaces that are present in many applied problems, especially those found in industrial endeavors.

At the printing of this volume, a number of other volumes are already in progress. Volume 3 will focus on seismic exploration, and Volume 4 on emerging opportunities related to parallel computing. Other topics in scientific computing are among those currently under consideration for future volumes. Members of a rotating editorial board will encourage active participation of the mathematical and scientific community in selection of topics for future volumes.

The series is being launched with several volumes in "energy mathematics" and the editorial board gratefully acknowledges the faculty and administration of the University of Wyoming for their cooperation in connection with the Special Year on Energy-Related Mathematics held in Laramie during the 1982–83 academic year. A great deal of the planning of these first volumes was facilitated through contact with visitors to the special program at the University of Wyoming.

H. T. BANKS

PREFACE

Combustion is, to a large extent, a superset of fluid mechanics, but—unlike that subject—one does not find within it a rich interaction of mathematics and experiment. Without offending those (or their memory) who have made important theoretical contributions to the subject, it can safely be said that combustion has been (and remains) primarily an empirical science. As evidence (at least for the Western world) one need only examine the proceedings of the biennial International Symposia organized by the Combustion Institute.

When theoretical work has been carried out, it has seldom exploited the full range of mathematical tools familiar in mechanics. Engineering approximations of an ad hoc nature are often made at different stages of the analysis, which obscures the theoretical model, and therefore the conclusions that may be drawn from its success or failure in emulating the physical world. Preoccupation with "reality" (as the physical world is called) leads to resistance—even rejection—of models that do not parallel the physical world. The notion of a mathematical universe in which one can set aside density variations in order to more readily explore the interaction of other mechanisms; a universe in which, having defined a model, one can construct a body of theoretical knowledge without further reference to the physics; a universe which will, provided the model has been judiciously chosen, emulate the physical world *here and there* and thus provide vivid insights into that world; such a notion is not one that many in combustion are comfortable with. Experimental results are always of value, but mathematical conclusions tend to be subjected to the test of immediate physical relevance. In short, the world of combustion is not one in which the goals and methods of applied mathematics are widely understood.

Fortunately, this situation is rapidly changing. In recent years applied mathematicians have been attracted in growing numbers to a subject whose richness affords ample scope for their talents. Asymptotic methods and bifurcation theory have made a significant impact in the last decade. And the stunning success of constant density, small heat release models in unravelling the mysteries of deflagration instabilities, together with other important contributions, is convincing an ever widening circle of combustion experts that mathematics can play a vital role.

Mathematics is having a new impact on combustion, and it is this, rather than the development of new mathematics, that makes the subject a "frontier", suitable for inclusion in this series.

This volume contains five chapters, each of which deals with topics in which mathematics plays an important role. Only in a partial sense is it a self-contained exposition of the subject; it is difficult to make useful contributions to combustion without undergoing an extensive apprenticeship in the relevant physics, and for this the reader must turn elsewhere. The well-known book by B. Lewis and G. von Elbe, cited in Chapter I, is an excellent source, as are graduate college texts such as *Combustion Fundamentals* by R. Strehlow.

The first chapter of this volume introduces the equations of the subject, and briefly discusses such fundamental matters as the small Mach number approximation, deflagration waves, kinetic modelling, ignition, the hydrodynamic model, and activation energy asymptotics. Some specific problems are described in order to convey something of the flavor of the theoretical questions that have been considered in recent years.

The remaining chapters deal with more specific questions. Combustion modelling, especially the kinetics, is a difficult subject often complicated by poorly known data. How this affects the accuracy and reliability of the results is the subject of sensitivity analysis, discussed in Chapter II. This is not a tool unique to combustion theory of course, and at the present time has had a greater impact outside of combustion than within it. Many would agree with the suggestion that this should change.

Chapter III discusses turbulence. Although laminar flames are not a mathematical or laboratory artifice, turbulent combustion plays a vital role in many situations of great practical significance. Interestingly enough, turbulent combustion is not simply classical turbulence plus the complication of exothermic reaction. It can, instead, be less complicated; unique simplifications give it a flavor all its own.

Chapters IV and V deal with compressible flows. Detonation is one of the oldest topics in the field, and one of the most difficult. The analysis of model equations, simpler than the exact ones, is useful in unravelling these difficulties, and such an approach forms the heart of Chapter IV. Chapter V discusses the behavior of finite amplitude waves in combustible gases. The material in these two chapters is relevant to the important and poorly understood subject of deflagration-to-detonation transition.

JOHN D. BUCKMASTER

THE MATHEMATICS OF COMBUSTION

CHAPTER I

An Introduction to Combustion Theory[1]

JOHN D. BUCKMASTER

1. The fundamental equations. The equations that govern the motion of a chemically reacting gas are complex. Some of this complexity is essential: the heat released by exothermic reaction is what makes a flame hot. Much is not: that each component of the reacting mixture has a different specific heat is a quantitative detail, usually swamped by other uncertainties. Thus, in our discussion of the equations governing combustion, we shall deal only with those ingredients that are usually important, omitting those that are likely to be important only in special cases.

The simplest way to do this is in the context of a dilute mixture, one whose fundamental physical properties (viscosity, thermal conductivity, etc.) are defined by a single component. In air mixtures, nitrogen might play this role. Mixtures which are not dilute can often be modelled by the same equations if appropriate averages are adopted for the various properties. The question as to when and how a mixture can be represented by "single-fluid" equations is an important one, but is not discussed in these pages.

We shall start by writing down that subset of the equations that will be so familiar to the reader with a passing acquaintance of fluid mechanics that it needs little comment. This consists of the equation of mass conservation,

$$(1) \qquad \frac{\partial \rho}{\partial t} + \nabla \cdot (\rho \mathbf{v}) = 0;$$

that of momentum,

$$(2) \qquad \rho \frac{D\mathbf{v}}{Dt} = \nabla \cdot \mathbf{\Sigma}$$

where $\mathbf{\Sigma}$, the stress tensor, is given by

$$(3) \qquad \mathbf{\Sigma} = -(p + \tfrac{2}{3}\kappa \nabla \cdot \mathbf{v})\mathbf{I} + \kappa[\nabla \mathbf{v} + (\nabla \mathbf{v})^T]$$

---

[1] *Acknowledgment.* The author was supported by the National Science Foundation and the Army Research Office during the time in which this chapter was written, and the editing chores for the volume were undertaken.

for a Newtonian fluid with no bulk viscosity; and the equation of state,

$$p = R\rho T, \quad R = \frac{\mathcal{R}}{\mathrm{MW}}. \tag{4}$$

The molecular weight is that of the diluent or, for a nondilute mixture, may be defined by

$$\mathrm{MW} = \sum n_i \, \mathrm{MW}_i = [\sum Y_i \, \mathrm{MW}_i^{-1}]^{-1}, \tag{5}$$

where n_i is the number fraction of the ith component, Y_i the mass fraction.

The energy equation deserves a more detailed treatment since some of its ingredients might be unfamiliar to the combustion neophyte. Consider a fluid in which no irreversible processes occur. The energy density is

$$E_d = \tfrac{1}{2}\rho v^2 + \rho u, \tag{6}$$

and the energy flux density is

$$\mathbf{E}_f = \rho\mathbf{v}(\tfrac{1}{2}v^2 + h) \tag{7}$$

so that

$$\frac{\partial E_d}{\partial t} + \nabla \cdot (\mathbf{E}_f) = 0. \tag{8}$$

Indeed, the formula (7) is usually established by showing that (8) is true by using the earlier equations.

Since

$$\mathbf{E}_f = \rho\mathbf{v}(\tfrac{1}{2}v^2 + u) + p\mathbf{v},$$

the term $\nabla \cdot (p\mathbf{v})$ in (8), when integrated over a volume with boundary S, becomes

$$\iint_S p v_n \, dS$$

and can be thought of as pressure work on the fluid within the volume. Using mass conservation, (8) can be written in the form

$$-\frac{\partial p}{\partial t} + \rho \frac{D}{Dt}(\tfrac{1}{2}v^2 + h) = 0. \tag{9}$$

There are two ways in which this last equation is modified when irreversible phenomena are included. The boundary work term now includes viscous stresses so that $p\mathbf{v}$ is replaced by $-\mathbf{v} \cdot \boldsymbol{\Sigma}$. And two additional fluxes arise, the Newtonian heat flux, and the enthalpy flux associated with the diffusion fluxes. These are

(10) $$-\lambda \nabla T + \rho \sum_{1}^{N} Y_i h_i \mathbf{V}_i.$$

The diffusion velocities $\{\mathbf{V}_i\}$ depend on the various concentration gradients. For a dilute gas it is appropriate to adopt a Fickian law relating $\mathbf{V}_i$ to the gradient of Y_i for all components but the diluent (the Nth component), so that

(11) $$\rho Y_i \mathbf{V}_i = -\mu_{ii} \nabla Y_i, \quad i \neq N.$$

Since, by definition,

(12) $$\sum_{1}^{N} Y_i \mathbf{V}_i = 0,$$

we have

(13) $$\rho Y_N \mathbf{V}_N = \sum_{1}^{N-1} \mu_{ii} \nabla Y_i.$$

The separate enthalpies can be written as

(14) $$h_i = h_i^0 + \int_{T_0}^{T} C_{p_i} \, dT$$

where C_{p_i} is the specific heat and h_i^0 is the heat of formation of the ith component at the temperature T_0.[2]

It follows that the diffusion enthalpy flux, the second term in (10), is

$$\begin{aligned}
\rho \sum_{1}^{N} Y_i h_i \mathbf{V}_i &= \rho \sum_{1}^{N} Y_i h_i^0 \mathbf{V}_i + \rho \sum_{1}^{N} Y_i \mathbf{V}_i \int_{T_0}^{T} C_{p_i} \, dT \\
&= \rho \sum_{1}^{N} Y_i h_i^0 \mathbf{V}_i - \rho \sum_{1}^{N-1} Y_i \mathbf{V}_i \int_{T_0}^{T} (C_{p_N} - C_{p_i}) \, dT \\
&\simeq \rho \sum_{1}^{N} Y_i h_i^0 \mathbf{V}_i
\end{aligned}$$

(15)

---

[2] T_0 is usually taken to be 298°K and the h_i^0 for elemental gases in their natural molecular state at that temperature (e.g. H_2, O_2) are zero. Thus for a constant pressure process represented by

$$H_2 + \tfrac{1}{2} O_2 \rightarrow H_2O \quad \text{(gas)},$$

for which the initial and final temperatures are 298°K, the enthalpy of the initial mixture is zero, that of the water -57.8 kcal/mole [1]. Energy equal to 57.8 kcal must be removed for every mole of water generated, in order to prevent the exothermic reaction from creating a temperature rise.

if we neglect the second term (heats of formation are large compared to differences in specific heats); this equals

$$\text{(16)} \qquad \sum_{1}^{N} \mu_{ii}(h_N^0 - h_i^0)\nabla Y_i.$$

Thus the formula (7) for the energy flux must be replaced by

$$\text{(17)} \qquad E_f = \rho\mathbf{v}(\tfrac{1}{2}v^2 + h) - \mathbf{v}\cdot\mathbf{\Sigma}' - \lambda\nabla T + \sum_{1}^{N} \mu_{ii}(h_N^0 - h_i^0)\nabla Y_i$$

where

$$\text{(18)} \qquad \mathbf{\Sigma}' = \mathbf{\Sigma} + p\mathbf{I}, \qquad h = \sum_{1}^{N} Y_i h_i^0 + \int_{T_0}^{T} C_p\, dT,$$

and (9) is replaced by

$$\text{(19)} \qquad -\frac{\partial p}{\partial t} + \rho\frac{D}{Dt}(\tfrac{1}{2}v^2 + h) - \nabla\cdot(\mathbf{v}\cdot\mathbf{\Sigma}')$$

$$= \nabla\cdot(\lambda\nabla T) + \sum_{1}^{N} (h_i^0 - h_N^0)\nabla\cdot(\mu_{ii}\nabla Y_i).$$

For a dilute mixture C_p depends only on the temperature and not on the mixture composition.

Each component of the mixture, represented by the mass fraction Y_i, satisfies a conservation equation of the form

$$\text{(20)} \qquad \rho\frac{DY_i}{Dt} - \nabla\cdot(\mu_{ii}\nabla Y_i) = \dot\rho_i, \qquad i = 1,\cdots, N-1,$$

$$Y_N = 1 - \sum_{1}^{N-1} Y_i,$$

where $\dot\rho_i$ represents the mass increase per unit volume of the ith species due to chemical reaction. Since the left side of (19) contains a term $\rho \sum_1^N h_i^0\, DY_i/Dt$ we can use (20) to rewrite (19) in the form

$$\text{(21)} \qquad -\frac{\partial p}{\partial t} + \rho\frac{D}{Dt}\left(\tfrac{1}{2}v^2 + \int_{T_0}^{T} C_p\, dT\right) - \nabla\cdot(\mathbf{v}\cdot\mathbf{\Sigma}')$$

$$= \nabla\cdot(\lambda\nabla T) - \sum_{1}^{N} h_i^0\dot\rho_i,$$

where we have used the condition $\sum_1^N \dot\rho_i = 0$. This is identical to a single-fluid equation with the heat released by reaction appearing as a source term.

Equations (1), (2), (4), (20) and (21) are adequate to describe most laminar flames, and are an appropriate starting point for turbulence modelling (Chapter III). Their description is completed by a discussion of the reaction terms.

2. Chemical kinetics. The quantity $\dot{\rho}_i$ is the mass of the ith component generated in unit time per unit volume due to chemical reaction. Usually there will be many reactions occurring simultaneously. For example, Clarke [2] has used the following system to describe the steady hydrogen/oxygen diffusion flame:

(22)
$$H_2 + M \rightleftarrows 2H + M$$
$$H_2 + OH \rightleftarrows H_2O + H$$
$$H + O_2 \rightleftarrows OH + O$$
$$O + H_2 \rightleftarrows OH + H$$
$$OH + H + M \rightleftarrows H_2O + M;$$

here M represents all species.

For 2-body reactions the reaction rate depends on the energy of collision. Consider a Maxwellian distribution of molecules each of mass m; the number N with speed between c_0 and infinity has a temperature dependence

(23) $\quad N \propto T^{-3/2} \int_{c_0}^{\infty} \exp\left(-\frac{c^2 m}{2kT}\right) dc \quad (k = \text{Boltzmann's constant}).$

If only high energy collisions can trigger chemical change, the number of molecules that can effect this change can be estimated from the asymptotic behavior of (23) for large c_0; we have

(24) $\quad N \propto T^{-1/2} \exp\left(-\frac{mc_0^2}{2kT}\right) = T^{-1/2} \exp(-E/RT)$

where

(25) $$E = c_0^2 \frac{mR}{2k}$$

has dimension (velocity)^2 and is called the activation energy. The reaction rate is proportional to this number of sufficiently energetic molecules and so will have an exponential dependence on temperature. This is called the Arrhenius law and plays an important role in combustion theory.

A strong temperature dependence is also a feature of decomposition, e.g. dissociation. This occurs when a significant number of the molecules have high vibrational energies. For third order reactions such as the reverse reaction (22a) or the forward reaction (22e), the temperature dependence is weakly algebraic.

Each of the reactions shown in (22) is reversible. When the forward rate is equal to the reverse rate the reaction is said to be in equilibrium. Consider (22a) for example,

$$\text{(26)} \qquad H_2 + M \underset{k_r}{\overset{k_f}{\rightleftarrows}} 2H + M.$$

According to the law of mass action the reaction rate is proportional to the number density of each reactant. Thus

$$\text{(27)} \qquad k_f \propto n_{H_2} n_M \propto \rho^2 Y_{H_2} Y_M,$$
$$k_r \propto n_H^2 n_M \propto \rho^3 Y_H^2 Y_M,$$

where temperature dependent proportionality factors are not shown. At equilibrium k_f and k_r are equal so that

$$\text{(28)} \qquad \frac{Y_{H_2}}{\rho Y_H^2} = K(T)$$

where K, although temperature dependent, is called the equilibrium constant. The full form for k_f is usually taken to be

$$\text{(29)} \qquad k_f = B\rho^2 Y_{H_2} T^\alpha \exp(-E/RT)$$

with a similar form for k_r (note that $Y_M = 1$ since M represents all species).

Contributions to $\dot{\rho}_i$ come from every reaction in which the ith species participates, either as a reactant or a product. The rates must be weighted according to the stoichiometric coefficients and the molecular weights. Thus if the reaction

$$H_2 + \tfrac{1}{2}O_2 \to H_2O$$

consumes H_2 at a rate of k molecules per unit volume per unit time, and is the only one in which H_2 and O_2 participate, we have

$$\text{(30)} \qquad \dot{\rho}_{O_2} = -m_{H_2}k, \qquad \dot{\rho}_{O_2} = -\tfrac{1}{2}m_{O_2}k.$$

It is apparent that a system such as (22), described by rates of the form (29), adds a great deal of complexity to the mathematical description. Only under exceptional circumstances is an analytical treatment possible. For this reason simplified kinetic models are often adopted. Consider, for example, the combustion of methane in oxygen. The essential nature of this process is that methane and oxygen are consumed, and products and heat are generated. It can, therefore, be modelled by a one-step irreversible process, represented by

$$\text{(31)} \qquad CH_4 + 2O_2 \overset{k}{\to} CO_2 + 2H_2O,$$

with

(32) $$k = B\rho^n Y_{CH_4}^{\gamma_1} Y_{O_2}^{\gamma_2} T^\alpha \exp(-E/RT).$$

The various parameters must be determined empirically; in particular γ_1 and γ_2 have no connection with the stoichiometric coefficients 1 and 2 that would be appropriate if (31) were a true reaction rather than a representation of many.

An even simpler model can be appropriate if the reactants are supplied as a homogeneous mixture. In general one of the reactants (the deficient one) will be present in less than stoichiometric proportion and so will be completely consumed along with a portion of the other (the surplus reactant). If the mass fraction of the deficient component is Y, the reaction can be represented by

(33) $$Y \rightarrow \text{products}$$

at a rate

(34) $$k = B\rho^n Y^\gamma T^\alpha \exp(-E/RT).$$

The simple models represented by (31) and (33) play a central role in much of the mathematical treatment of combustion. Whether they are appropriate or not depends, to a large extent, on the aim of the analysis. We shall return to this question later, after some elementary combustion concepts have been introduced. For the moment we shall proceed with our discussion using a one-step irreversible model when an explicit choice for the kinetics must be made.

3. The small Mach number approximation. For many flames, velocities are small (say 100 cm/s), temperatures large (2000–3000°F). A representative Mach number Ma is then small and this leads to important simplifications of the governing equations. Except for detonation, the approach to it, or acoustic propagation, the simplified set is at the heart of all combustion modelling.

At small Mach numbers the kinetic energy is small in comparison with the thermal energy, and so can be neglected in the energy equation (21). The viscous term $\nabla \cdot (\mathbf{v} \cdot \mathbf{\Sigma}')$ is likewise negligible, depending as it does on the square of the velocity. Equation (21) can therefore be replaced by

(35) $$-\frac{\partial p}{\partial t} + \rho C_p \frac{DT}{Dt} = \nabla \cdot (\lambda \nabla T) - \sum_1^N h_i^0 \dot{\rho}_i.$$

Moreover, from the momentum equation, spatial changes in p satisfy the estimate

$$(\delta p)_s \sim \rho v^2$$

whence

(36) $$\frac{(\delta p)_s}{p} \sim \text{Ma}^2.$$

Thus p has the representation

(37) $$p = p_0(t) + \text{Ma}^2 p_1(x, y, z, t) + o(\text{Ma}^2)$$

and it is p_0 that appears in the equation of state, the simplified energy equation and the reaction terms, p_1 in the momentum equation (the small pressure gradient drives the small velocities).

With two variables (p_0, p_1) replacing one (p) it is not clear that the reduced system is complete. There is no difficulty, however: for an unbounded problem $p_0(t)$ is usually an assigned quantity; for a bounded problem global considerations, manifest through an application of Gauss's law, relate changes in p_0 to the chemical activity within the volume, and the thermal fluxes at the boundary. Consider the case when C_p is constant. Noting that

(38) $$\rho \frac{DT}{Dt} = \frac{\partial}{\partial t}(\rho T) + \nabla \cdot (\rho v T) = \frac{d}{dt}\left(\frac{p_0}{R}\right) + \nabla \cdot (\rho v T),$$

(35) is

(39) $$\left(\frac{C_p}{R} - 1\right)\frac{dp_0}{dt} + C_p \nabla \cdot (\rho v T) = \nabla \cdot (\lambda \nabla T) - \sum_{1}^{N} h_i^0 \dot{\rho}_i,$$

which, upon integration over the entire combustion field of volume V_0, yields the compatibility condition

(40) $$V_0\left(\frac{C_p}{R} - 1\right)\frac{dp_0}{dt}$$
$$= -C_p \iint_{\partial V_0} \rho v_n T \, dS + \iint_{\partial V_0} \lambda \frac{\partial T}{\partial n} \, dS - \iiint_{V_0} \sum_{1}^{N} h_i^0 \dot{\rho}_i \, dV.$$

The initial value problem for the small Mach number system has some peculiarities, which will be identified in §12.

4. The plane flame or deflagration wave. We now turn our attention to the formulation of a central problem in combustion, the so-called deflagration wave. Consider a homogeneous mixture of fuel and oxidizer. This can support a one-dimensional premixed flame which is wavelike in the sense that it can propagate at constant speed with unchanging form relative to the mixture. The gas ahead of the wave is cold and fresh; behind it is hot and burnt.

We shall adopt a coordinate frame fixed to the flame with the flow passing at speed u at right angles to the flame in the direction of increasing x. If the subscript f is used to denote the unburnt state, corresponding to $x \to -\infty$, u_f is the speed of the wave relative to the fresh mixture. This is the flame speed, to be calculated as part of the solution.

For this steady one-dimensional problem mass conservation implies

(41) $$\rho u = \rho_f u_f = M.$$

With one-step kinetics ($N = 2$) the energy and species equations then have the form

(42) $$MC_p \frac{dT}{dx} = \frac{d}{dx}\left(\lambda \frac{dT}{dx}\right) - h_1^0 \dot{\rho}_1,$$

$$M \frac{dY_1}{dx} = \frac{d}{dx}\left(\mu_{11} \frac{dY_1}{dx}\right) + \dot{\rho}_1$$

where $\dot{\rho}_1$ is a function of ρ, Y_1 and T (the density is related to the temperature by the equation of state (4) where p is the constant p_0). A linear combination of equations (42) can be formed that contains no reaction term,

(43) $$\frac{MC_p}{h_1^0} \frac{dT}{dx} + M \frac{dY_1}{dx} = \frac{1}{h_1^0} \frac{d}{dx}\left(\lambda \frac{dT}{dx}\right) + \frac{d}{dx}\left(\mu_{11} \frac{dY_1}{dx}\right).$$

A key feature of the one-step kinetic scheme represented by (31), (33) is that reaction can only cease when the reactant is completely consumed; this corresponds to the equilibrium state. Thus Y vanishes in the burnt gas ($x \to +\infty$).[3]

With these facts in mind, integration of (43) from ($-\infty$) to ($+\infty$) yields a value for T_b, the temperature of the burnt gas,

(44) $$T_b = T_f + \frac{h_1^0}{C_p} Y_{1f};$$

this is the adiabatic flame temperature.

Equations (42) can be solved numerically. A convenient nondimensional system can be defined by using the characteristic length λ/MC_p,[4] characteristic temperature h_1^0/C_p, and then we have

(45) $$\frac{dT}{dx} = \frac{d^2T}{dx^2} + \frac{D}{M^2} Y^\gamma T^\beta \exp(-\theta/T),$$

$$\frac{dY}{dx} = \frac{1}{\text{Le}} \frac{d^2Y_1}{dx^2} - \frac{D}{M^2} Y_1^\gamma t^\beta \exp(-\theta/T).$$

---

[3] Reaction must also vanish far upstream, which is possible only if (34) is modified by introducing a cut-off temperature T_c ($> T_f$) below which all reaction ceases. The difficulty which forces this modification is known as the cold-boundary difficulty, and has been extensively discussed in the literature.

[4] λ and μ are usually temperature dependent, but here we approximate them by constants.

Here Le $= \lambda/C_p\mu_{11}$ is the Lewis number, θ is a nondimensional activation energy, and D is proportional to B and, in addition, depends on p_0; T and x are now dimensionless variables. Typical solutions are shown in Fig. 1. Later we shall discuss an analytical approach but before that we want to return to the matter of kinetic modelling.

5. Kinetics revisited. Whether or not the simple kinetics embodied in (33) is good enough depends on the goal of the analysis. The parameters at our disposal can always be chosen so that the theoretical flame speed coincides with the experimental value for specific upstream conditions; this choice is not unique. The behavior of small perturbations, as in a stability analysis say, does not differ qualitatively with different choices of γ and β. Thus the physical mechanisms responsible for certain instabilities of the deflagration wave can be satisfactorily explored by assigning convenient values to these parameters (1 and 0, for example). Variations of flame speed with mixture strength for fuel-lean hydrocarbon flames can be adequately fitted using (33).

On the other hand, for fuel-rich hydrocarbon flames, (33) is inadequate for predicting flame-speed variations. The two-component model (31) is satisfactory for this purpose provided γ_1 and γ_2 are carefully chosen [3], but this success is misleading since variations of flame temperature are inadequately described. As an example, consider the combustion of methane in air, represented by (31). A stoichiometric mixture contains, by weight, four times more oxygen than methane. For a lean mixture, equilibrium corresponds to the complete consumption of methane, so that a formula like (44) is valid and, assuming air is 21% by volume O_2, this has the form

$$(46) \qquad T_b = \left[\frac{.008 + [CH_4]}{.063}\right] \cdot (2250)°K$$

where $[CH_4]$ is the mass fraction of methane. This describes the transition from a temperature of 293°K (room temperature) when $[CH_4]$ vanishes, to

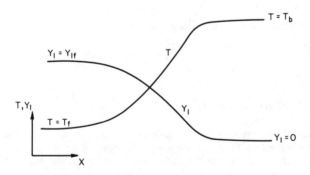

FIG. 1. *Structure of the deflagration wave.*

the experimentally observed maximum of 2250°K at stoichiometry when [CH$_4$] = .055. In practice there is a limit to how cool the flame can be; too lean a mixture gives too cool a flame, which cannot survive. This lean limit occurs at [CH$_4$] = .03, at a temperature predicted by (46) of 1350°K, 900°K below the maximum. The measured limit temperature is 1500°K, 750°K below the maximum, so that the error in using (46) is reasonably small.

On the rich side of stoichiometry, equilibrium for our irreversible model corresponds to the complete consumption of oxygen, and (46) is replaced by

$$(47) \qquad T_b = \left[\frac{.088 + .25[O_2]}{.063} \right] (2250)°K$$

where, at stoichiometry, [O_2] = .22. The rich limit corresponds to [O_2] = .21 for which (47) predicts a flame temperature of 2180°K compared to the experimental value of 1900°K. Here the model predicts unrealistically small variations in temperature with mixture strength.

Success of the model on the lean side can be attributed to the fact that complete oxidation occurs (there is a surplus of oxygen) so that (31) reasonably approximates the initial and final states. Failure on the rich side is due to the fact that the oxidation is incomplete, CO and OH are present, and the temperature depends on the equilibrium of reversible reactions. Equation (31) is an inadequate representation.

If this is important a more sophisticated model is required, such as [4]

$$(48) \qquad \begin{array}{c} X_n + nO \to nXO \\ 2XO + O_2 \rightleftarrows 2XO_2 \end{array}$$

where X represents the atoms C and H and the second (reversible) reaction is assumed to be in equilibrium.

6. Kinetic modelling in ignition.

This matter of kinetic modelling also arises when we consider the question of auto-ignition, the genesis and evolution of vigorous burning in a combustible material. Auto-ignition is a self-accelerating process and one important mechanism is adequately described by one-step Arrhenius kinetics. To demonstrate this, consider the spatially homogeneous energy equation

$$(49) \qquad \rho C_p \frac{dT}{dt} = -h_1^0 \dot{\rho}_1$$

with initial condition $T(0) = T_0$.

If we ignore density and concentration changes (a realistic assumption during the early stages of the ignition process) and assume no algebraic dependence of $\dot{\rho}_1$ on T (for convenience only) then, after appropriate scalings, (49) can be written in the nondimensional form

$$\text{(50)} \qquad \frac{dT}{dt} = \frac{1}{\theta}\exp\left(\theta - \frac{\theta}{T}\right), \qquad T(0) = 1,$$

where, it will be recalled, θ is a nondimensional activation energy.

The heat generated by reaction raises the temperature, which increases the vigor of the reaction—this is the process of self-acceleration. It is most clearly seen when θ is very large, a characteristic of combustion. For then, at early times, we can seek a solution to (50) of the form

$$\text{(51)} \qquad T \sim 1 + \frac{1}{\theta} T_1 + o\left(\frac{1}{\theta}\right)$$

whence

$$\text{(52)} \qquad \frac{dT_1}{dt} = \exp(T_1), \qquad T_1(0) = 0,$$

with solution

$$\text{(53)} \qquad T_1 = -\ln(1 - t).$$

T_1 is unbounded as $t \to 1$, implying a breakdown in the representation (51). Self-acceleration becomes so strong that the temperature deviates significantly and rapidly from the initial value.[5]

The process that we have just described is known as thermal ignition. A quite different mechanism is responsible for the ignition of hydrogen at (say) 400°C, 4mm/Hg pressure. In this case the kinetics can be described by the following system;

(54) (i) $\qquad H_2 + M \to 2H + M$

(ii) $\qquad H + O_2 \to OH + O$
$\qquad\qquad O + H_2 \to OH + H$

(iii) $\qquad OH + H_2 \to H_2O + H$

(iv) $\qquad H + H + M \to H_2 + M;$

M represents all species. Overall, hydrogen and oxygen burn to form water.

The reactions have been placed in one of four categories. The first of these, (i), is called *initiation*; collision of molecular hydrogen with any other

---

[5] In an inhomogeneous problem this runaway occurs in a small region and ultimately leads to a deflagration wave travelling from the ignition point out into the unburned gas. A complete treatment, valid in the limit $\theta \to \infty$, is given by Buckmaster and Ludford [5] following work of Kassoy and (independently) Kapila. Corrections which are mathematically important, but do not affect the fundamental picture, have been described in [6].

body generates atomic hydrogen. Quantitatively this is not a significant reaction but it plays a crucial role in providing the seeds for the subsequent reactions.

The second category contains the *branching* reactions. Each atom of hydrogen or oxygen (the active bodies) breaks up one molecule, at the same time generating two active bodies (OH, H or O) which, in turn, break up two molecules to form three active bodies, ... This is a rapidly accelerating process, responsible for ignition.

The third category is *propagation*; the active body OH does its job and produces a single active body.

The fourth and final category is *termination*; H atoms are removed. Termination can also occur when H atoms collide at a wall. If termination is strong enough it can overcome the branching, and ignition will not occur.

In contrast to thermal ignition, the initial process here is essentially isothermal since the branching/propagation reactions are weakly exothermic. The high temperatures eventually achieved come from the highly exothermic recombination reaction (iv), which must eventually limit the number of atomic particles that are generated.

A system such as (54) can be modelled in a way proposed by Zeldovich:

(55) $$A + B \to 2B$$
$$2B \to C.$$

The first reaction has a high activation energy and no heat release, the second has no temperature dependence, large heat release. For the ignition problem a nonzero value of B must be specified initially as a substitute for the initiation or seeding reaction. Thus, the ignition problem for this model can be described, following Kapila [7], by the equations

(56) $$\frac{dY_A}{dt} = -\Omega, \quad \frac{dY_B}{dt} = \Omega - Y_B, \quad \frac{dT}{dt} = Y_B,$$
$$\Omega = DY_A Y_B \exp\left(\theta - \frac{\theta}{T}\right),$$
$$T(0) = 1, \quad Y_A(0) = \alpha, \quad Y_B(0) = \beta/\theta;$$

this system is designed with the limit $\theta \to \infty$ in mind.

The early development, analogous to (51) for the thermal model, is described by

(57) $$T = 1 + \frac{1}{\theta}T_1 + \cdots, \quad Y_A = \alpha\left(1 + \frac{1}{\theta}Y_{A_1} + \ldots\right), \quad Y_B = \frac{1}{\theta}Y_{B_1} + \cdots.$$

The nature of the solution depends on the value of D. If D is small, termination dominates; if D is large, branching dominates and ignition occurs. The critical value of D is the smallest root of

(58) $$\exp(\alpha D) = \alpha D \exp(1 + \beta)$$

and for D greater than the critical T_1, Y_{A_1} and Y_{B_1} all become singular at some critical time t_c. T_1 is logarithmically singular as in (53), but Y_{A_1}, Y_{B_1} are much larger, behaving like $(t_c - t)^{-1}$. This signals the onset of ignition as a process characterized in its early phases by $O(1)$ changes in Y_A and Y_B, small changes in T.

Just because a branching model must be used for a particular gas in order to realistically simulate the ignition process, does not mean that a thermal model is inappropriate to describe the high temperature reactions sustaining deflagration. Different conditions can require different models, since reactions that are important under some circumstances can be unimportant under others. To demonstrate this, consider the Zeldovich model (55) described by

(59) $$\dot{\rho}_A = -D_1 Y_A Y_B e^{-\theta/T}, \qquad \dot{\rho}_B = D_1 Y_A Y_B e^{-\theta/T} - D_2 Y_B^2;$$

these formulas are slightly different from (56). If $D_2 \gg D_1 e^{-\theta/T}$ the concentration of reactant B will be very small; any generated by the first reaction is rapidly consumed by the second. Indeed there will be a reactive balance obtained formally by setting $\dot{\rho}_B \sim 0$ so that

(60) $$D_1 Y_A Y_B e^{-\theta/T} \sim D_2 Y_B^2$$

and this determines Y_B. Thus

(61) $$\dot{\rho}_A \sim -\frac{D_1^2}{D_2} Y_A^2 \exp\left(-\frac{2\theta}{T}\right)$$

and the heat release, proportional to $D_2 Y_B^2$, is proportional to

(62) $$Y_A^2 \exp\left(-\frac{2\theta}{T}\right).$$

The results (61), (62) are equivalent to a one-step process involving reactant A alone.

Peters [8] has carried out numerical calculations for a deflagration wave approaching a cold wall with $\theta = 20,000$, T_b (the adiabatic flame temperature) equal to 2000. He found that for $D_2/D_1 = 10^{-2}$ the results are as for a one-step reaction; for $D_2/D_1 = 10^{-8}$ there are significant differences.

To conclude our brief and incomplete discussion of kinetics, it is to be noted that many of the advances that have been made in combustion in the last ten years or so have been done in the context of one-step irreversible kinetics with large activation energy. Our understanding of ignition, flame stability, interaction between flames and flows, and diffusion flame structure, has increased significantly as a consequence of this work, but the vigor of such developments has declined recently—there is a limit to what can be done with but a simple model. Clearly there is a need to develop mathematical frameworks that can cope with more sophisticated kinetic modelling, and although there have been attempts of this type, much remains to be done.

A word of caution, however. Modelling should be done in a clear physical context as in the case of the system (48), and the Zeldovich system (55); otherwise the effort is likely to be nothing more than a mathematical exercise. In a field that is dominated by empiricism and ad hoc intuitive reasoning, it is an unfortunate fact that mathematical modelling has to continually prove its worth, and failure to attend to the physics reinforces the prejudices of those who find such proofs difficult to appreciate.

The kinetic modelling discussed here is necessarily simple, designed as a starting point for mathematical analysis. For problems in which detailed complex kinetics is to be incorporated into numerical calculations, different questions can arise. Often, reaction rates and their variations with temperature are not well known; orders of magnitude uncertainties are commonplace. What effect these uncertainties have on the results is the subject of sensitivity analysis, discussed by Rabitz in Chapter II. Sensitivity analysis can be important to the modeller by suggesting that certain reactions are unimportant and so can be neglected.

7. **Acoustics and the small Mach number approximation.** The small Mach number equations derived earlier preclude acoustic disturbances and yet their effect is a subject of great importance. Certain types of rocket motor instabilities are intimately associated with such interactions. Mathematically the problem is one of multiple scales; the length and velocity scales for the acoustic disturbance are much larger than those associated with the flame structure, and only by accounting for this can we incorporate acoustics into a small Mach number description.

Consider the one-dimensional equations in the form

(63)
$$\frac{\partial \rho}{\partial t} + \frac{\partial}{\partial x}(\rho v) = 0, \quad p = \rho PT,$$
$$\rho \frac{Dv}{Dt} = -\frac{\partial p}{\partial x} + \tfrac{4}{3}\frac{\partial}{\partial x}\left(\kappa \frac{\partial v}{\partial x}\right),$$
$$\rho T \frac{DS}{Dt} = \rho C_p \frac{DT}{Dt} - \frac{Dp}{dt} = \tfrac{4}{3}\kappa \left(\frac{\partial v}{\partial x}\right)^2 + \frac{\partial}{\partial x}\left(\lambda \frac{\partial T}{\partial x}\right) - \sum h_i^0 \dot{\rho}_i.$$

S is the entropy.[6]

---

[6] These are the equations needed to describe plane fast deflagrations and detonations. They are the point of departure for Fickett's discussion of detonations in Chapter IV.

Small perburbations about a uniform chemistry-free state satisfy the equations

(64)
$$\frac{\partial \rho'}{\partial t} + \rho \frac{\partial v'}{\partial x} + v \frac{\partial \rho'}{\partial x} = 0, \qquad p' = \rho' RT + \rho RT',$$
$$\rho \frac{Dv'}{Dt} = \rho \frac{\partial v'}{\partial t} + \rho v \frac{\partial v'}{\partial x} = -\frac{\partial p'}{\partial x} + \tfrac{4}{3} \kappa \frac{\partial^2 v'}{\partial x^2},$$
$$\rho C_p \frac{DT'}{Dt} - \frac{Dp'}{Dt} = \lambda \frac{\partial^2 T'}{\partial x^2},$$

where ρ, v, p, T now represent undisturbed variables. To nondimensionalize these equations, we take as our reference quantities pressure p, temperature T, density ρ, velocity $c = \sqrt{\gamma RT}$ (the sound speed), time $\lambda/\rho C_p v^2$, and length $\lambda/\text{Ma}\, \rho C_p v$. Note that the characteristic time is the natural one, defined by parameters that control the flame structure, but the characteristic length is much larger than the flame length identified in §4, if Ma is small.

Equations (64) can now be written in nondimensional form:

(65)
$$\frac{\partial \rho'}{\partial t} + \text{Ma} \frac{\partial \rho'}{\partial x} + \frac{\partial v'}{\partial x} = 0, \qquad p' = \rho' + T',$$
$$\frac{Dv'}{Dt} = \frac{\partial v'}{\partial t} + \text{Ma} \frac{\partial v'}{\partial x} = -\frac{1}{\gamma}\frac{\partial p'}{\partial x} + \text{Ma}^2\, \tfrac{4}{3} \frac{\kappa C_p}{\lambda} \frac{\partial^2 v'}{\partial x^2},$$
$$\frac{DT'}{Dt} - \frac{R}{C_p}\frac{Dp'}{Dt} = \text{Ma}^2 \frac{\partial^2 T'}{\partial x^2}.$$

If terms of order $O(\text{Ma}^2)$ are neglected, these are the equations of acoustics. If on the other hand v is taken as the reference velocity rather than c, and $\lambda/\rho C_p v$ is chosen for the length, the limit equations are

(66)
$$\frac{\partial \rho'}{\partial t} + \frac{\partial \rho'}{\partial x} + \frac{\partial v'}{\partial x} = 0, \qquad p' = \rho' + T',$$
$$\frac{\partial p'}{\partial x} = 0,$$
$$\frac{\partial T'}{\partial t} + \frac{\partial T'}{\partial x} - \frac{R}{C_p}\frac{\partial p'}{\partial t} = \frac{\partial^2 T'}{\partial x^2}.$$

These are linearized versions of the small Mach number equations derived in §3.

In discussions of the interaction between acoustic waves and flames both scales might have to be considered. Waves propagating through flames are discussed in Chapter V.

8. The hydrodynamic model. The deflagration wave discussed in §4 has a thickness $\sim \lambda/MC_p$, a length also identified in §7; typically this is less than 1 millimeter. On a scale much larger than this the structure is unresolved and the flame appears as a discontinuity of temperature, density, and velocity. If the flame is curved it will generate a nonuniform flow field; this flow is incompressible when the Mach number is small. Thus it is necessary to solve Euler's equations for a constant density fluid, on each side of a front across which there are connection conditions representing the conservation of mass and momentum.

If V is the speed of the front back along its normal (Fig. 2), these connection conditions are

$$\rho_f(v_{n_f} + V) = \rho_b(v_{n_b} + V), \qquad v_{\perp f} = v_{\perp b},$$
$$p_f + \rho_f(v_{n_f} + V)^2 = p_b + \rho_b(v_{n_b} + V)^2. \tag{67}$$

T_b is the adiabatic flame temperature and $W \equiv v_{n_f} + V$ is the adiabatic flame speed, the speed of the front relative to the fresh gas. Both T_b and W are specified quantities. In the case of a stationary front ($V = 0$) we have (Fig. 3)

$$u_f = W, \qquad u_b = \sigma W \qquad (\sigma \equiv \rho_f/\rho_b),$$
$$v_f = v_b, \qquad p_b - p_f = \rho_f W^2(1 - \sigma). \tag{68}$$

The flow is refracted towards the normal on passage through the front.

The single most important application of the hydrodynamic model is the stability analysis done independently by Darrieus and Landau. They considered infinitesimal perturbations of the plane flame and showed that it is unconditionally unstable. The analysis will not be repeated here—it can be found in [9].

An elementary kinematic argument yields important insight into the manifestation of this instability (Fig. 4). A corrugated front traveling normal to itself at a uniform speed will develop sharp ridges in those portions of the front that are concave when viewed from the fresh gas. These ridges point towards the burnt gas.[7]

One of the most interesting and important demonstrations of this instability is seen when a nominally plane flame travels down a tube. Such flames are strongly curved and can be thought of as a single-celled manifestation of the hydrodynamic instability. This defines one of the few interesting problems in combustion that are purely hydrodynamical in nature, and we shall discuss

---

[7] A full understanding of the consequences can only come from a nonlinear analysis that goes beyond the hydrodynamical model and accounts for the effects of curvature on flame speed; such treatments have recently been reviewed by Sivashinsky [10]. Numerical solutions of elegant, well-motivated model equations show that the flame breaks up into cells delineated by sharp ridges. This type of structure is observed experimentally.

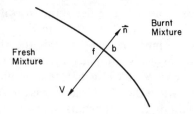

Fig. 2. *Hydrodynamic front.*

it briefly, as a vehicle for certain ideas that are essential ingredients of the hydrodynamical problem.

One effect of flame curvature is to increase the propagation speed down the tube. If this speed is V, fuel is consumed at a volumetric rate VA_t, where A_t is the cross-sectional area of the tube (Fig. 5). This equals the flux across the flame surface, WA_f, where A_f is the flame area, so that

$$(69) \qquad V = W\frac{A_f}{A_t} > W.$$

We have already noted that the flow is refracted on passage through the flame. In addition, a curved flame generates vorticity. In a frame moving with the flame so that $V = 0$ we can use the results (68) to calculate the total head (H) of the fluid behind the flame. This is given by the formula

$$(70) \quad H_b = p_b + \tfrac{1}{2}\rho_b(u_b^2 + v_b^2) = H_f + \tfrac{1}{2}\rho_f W^2(1 - \sigma) + \tfrac{1}{2}v_f^2(\rho_b - \rho_f).$$

The upstream vorticity can be assigned and here the natural choice is zero so that H_f is constant everywhere in the fresh mixture. H_b is constant on each streamline of the burnt gas, but will vary from streamline to streamline according to the variations of v_f along the flame front.

The generation of vorticity complicates the analysis. There have been numerical computations in which this vorticity is neglected, but this can only

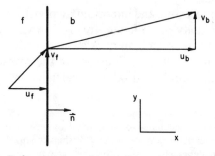

Fig. 3. *Refraction through a stationary hydrodynamic front.*

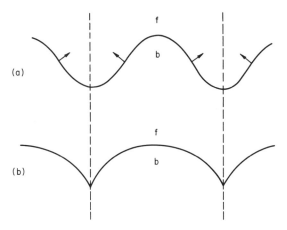

FIG. 4. *The evolution of ridges in a hydrodynamic front:* (a) *initial disturbance;* (b) *subsequent shape.*

be strictly justified for the special cases $\sigma = 1$ and $\sigma \to \infty$. The first choice is an ingredient of Sivashinsky's models (small heat release) but the second appears not to have been exploited. That it is true can be seen from (70), for, as $\sigma \to \infty$ with conditions in the fresh gas fixed, we have the order of magnitude estimates

(71) $$H_b \sim \rho_f W^2 \sigma, \qquad \delta(H_b) \sim \rho_f v_f^2,$$

whence

$$\frac{\delta(H_b)}{H_b} \sim \frac{1}{\sigma},$$

which is small. The velocity of the burnt gas is large $(O(\sigma W))$ and so its tangential component at the front, being $O(W)$, vanishes to leading order. The normal component is the constant σW and the leading order pressure is constant at the front also. A global momentum balance for the burnt gas then shows that, to leading order, the front must be plane; curvature, like the vorticity, is a perturbation for the tube problem.

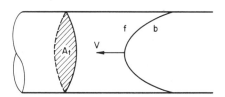

FIG. 5. *Propagation of a flame down a tube.*

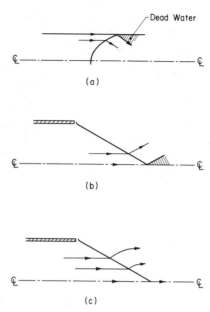

Fig. 6. *Resolution of the kinematic incompatibility:* (a) *flame in a tube;* (b) *flame tip with dead water;* (c) *flame tip with hydrodynamic singularity.*

The nature of these perturbations has not been investigated and it might be useful to do so since the nature of the flow is not clear in all respects. There is a difficulty at the point where the flame front intersects the wall. This intersection is not at right angles so that the refraction generates a kinematic incompatibility between the burnt gas and the wall, a difficulty recognized in [11]. Zeldovich [12] has proposed the existence of a dead water region between the limiting streamline and the wall (Fig. 6).

A similar difficulty occurs with plane flame tips, the plane version of the familiar Bunsen-burner flame (Fig. 6), in which the line of symmetry replaces the tube wall. In the context of *slender* tips, Buckmaster and Crowley [13] have described a solution without dead water in which the incompatibility is resolved through a singularity (in turn, assumed to be resolved on a scale small compared to the hydrodynamical one). This solution agrees qualitatively with observed flow fields.

The dead water hypothesis for the confined flame is an attractive one, but why it should be appropriate there but not for the burner flame is not clear. These hydrodynamic descriptions are not uniformly valid and we know little of the manner in which the nonuniformities should be resolved.

9. Activation energy asymptotics for premixed flames. In our discussion of the hydrodynamical model we assumed that the flame speed W is a specified quantity. It is to be calculated by solving the deflagration wave problem introduced in §4. In this, as in most combustion problems, the reaction plays a central role in the mathematical description. In the context of one-step irreversible kinetics the difficulties introduced thereby can be largely circumvented by an asymptotic treatment, valid in the limit of infinite activation energy. This approach has been widely discussed in recent years [5], [9], [10], [14], [15], so that there is little need for another review in these pages. But neither can we ignore it, since it has come to play such an important role in the theory.

The essential ideal of activation energy asymptotics is that, provided reactants are present, the reaction rate is a maximum at points of maximum temperature and is exponentially smaller (and so negligible) elsewhere—the Arrhenius (exponential) factor is extremely sensitive to temperature. In most contexts it then follows that reaction is confined to a thin zone called a flame sheet. For the premixed flame the thickness of this zone is given by the estimate $\lambda/M\ C_p \cdot 1/\theta$ where θ is a dimensionless activation energy E/RT_{ref}, where the reference temperature might be the adiabatic flame temperature. In this section we shall explore the consequences of this idea in the context of premixed flames.

On a scale defined by $\lambda/M\ C_p$ the reaction zone, in the limit, has a well defined normal. We shall examine the structure of a small portion of this sheet, choosing the x-axis to coincide locally with the normal. We then have, within the sheet, the leading order balance

$$(72) \quad 0 \sim \frac{d^2 T}{dx^2} + \mathcal{D} Y \exp(-\theta/T),$$

$$0 \sim \frac{1}{\text{Le}} \frac{d^2 Y}{dx^2} - \mathcal{D} Y \exp(-\theta/T), \quad \mathcal{D} = DM^{-2},$$

deduced from (45) when, for convenience, we choose $\gamma = 1$, $\beta = 0$. The sheet is thin and the reaction vigorous, hence the balance between diffusion and reaction terms.

It follows that

$$(73) \quad \frac{d^2}{dx^2}\left(T + \frac{1}{\text{Le}} Y\right) \sim 0$$

with solution

$$(74) \quad T + \frac{1}{\text{Le}} Y \sim C_1 x + C_2$$

where C_1 and C_2 are constants. This relation is valid throughout the reaction zone.

Now dT/dx can, in principle, assume $O(1)$ negative values immediately behind the flame sheet, but the corresponding flame structure is known to be unstable [16], so we shall suppose that dT/dx is $o(1)$ there. In the absence of a significant falling-off in temperature as the gas leaves the sheet, reaction can only cease when all the mixture is consumed, so that Y vanishes behind the sheet. It follows that C_1 vanishes and we shall define T_* so that (74) becomes

$$T + \frac{1}{\text{Le}} Y = T_*. \tag{75}$$

T_* is known as the flame temperature. Equation (72a) can now be written in the form

$$0 \sim \frac{d^2 T}{dx^2} + \mathcal{D} \, \text{Le}(T_* - T) \exp(-\theta/T). \tag{76}$$

When θ is very large, the Arrhenius factor is very small so that the balance expressed here can only make sense if $\mathcal{D}$ is correspondingly large. This motivates an examination of the distinguished limit defined by

$$\mathcal{D} = \hat{\mathcal{D}} \exp(\theta/T_*), \quad \theta \to \infty. \tag{77}$$

There is then a balance provided T differs by no more than an $O(\theta^{-1})$ amount from T_*, so that within the reaction zone we may write

$$T = T_* + (T - T_*), \quad |T - T_*| \ll T_*$$

and

$$\theta \left(\frac{1}{T_*} - \frac{1}{T} \right) \simeq \theta \frac{(T - T_*)}{T_*^2}$$

so that

$$0 \sim \frac{d^2}{dx^2} (T - T_*) - \text{Le} \, \hat{\mathcal{D}}(T - T_*) \exp\left[\theta \frac{(T - T_*)}{T_*^2} \right]. \tag{78}[8]$$

On the hot side of the flame sheet, $(T - T_*)$ and the temperature gradient both vanish, so that a single integration of (78) yields

$$0 \sim \frac{1}{2} \left[\frac{d}{dx} (T - T_*) \right]^2 - \text{Le} \, \hat{\mathcal{D}} \frac{T_*^2}{\theta} (T - T_*) \exp\left[\theta \frac{(T - T_*)}{T_*^2} \right]$$

$$+ \text{Le} \, \hat{\mathcal{D}} \frac{T_*^4}{\theta^2} \exp\left[\theta \frac{(T - T_*)}{T_*^2} \right] - \text{Le} \frac{\hat{\mathcal{D}} T_*^4}{\theta^2}. \tag{79}$$

---

[8] The reader who wishes to identify a formal asymptotic balance should note that $(T - T_*) = O(\theta^{-1})$, $d/dx = O(\theta)$ so that $\hat{\mathcal{D}}$ must be $O(\theta^2)$.

On the cold side of the sheet the temperature drops sharply, so that the exponentials in (79) vanish. It follows that the temperature gradient there is given by the formula

$$\frac{dT}{dx} = \sqrt{2 \, \text{Le} \, \hat{\mathcal{D}}} \cdot \frac{T_*^2}{\theta} = \sqrt{2 \, \text{Le} \, \mathcal{D}} \cdot \frac{T_*^2}{\theta} \exp(-\theta/2T_*). \tag{80}$$

On the scale defined by $\lambda/M\,C_p$, the reaction zone, in the limit, is a surface of discontinuity across which there is a jump in the gradients of T and Y. Thus the reaction term in (45) can be replaced by a Dirac δ-function of strength

$$\sqrt{2 \, \text{Le} \, \mathcal{D}} \cdot \frac{T_*^2}{\theta} \exp(-\theta/2T_*). \tag{81}$$

10. The δ-function model and its application. The formula (81) is an asymptotic result, valid in the limit $\theta \to \infty$. As such it should be used with appropriate asymptotic expansions beyond the flame sheet, matched with higher order descriptions within the flame sheet if necessary. The δ-function model abandons this framework and simply replaces the reaction term by the δ-function, with θ taken as a finite parameter (although simplifications valid when θ is large may be adopted in the subsequent analysis). Y is set equal to zero behind the flame sheet but no restriction is imposed on the temperature gradient there.

In view of the loose connection between the complex kinetics that properly describe real flames, and one-step irreversible kinetics, it is difficult to argue, a priori, that such a model is inferior to a treatment based on rational asymptotic analysis. It has the advantage of being much simpler—certain technical difficulties that can arise in the asymptotic treatment are irrelevant. And it can be understood without any grasp of asymptotic theory. It is, indeed, simple enough to be used as the foundation of an undergraduate textbook dealing with a great many important questions in flame theory, an opportunity that is yet to be exploited. We shall complete this section by applying the model to the deflagration wave of §4, and by a brief description of various problems in which either the model or its genesis has been used.

Deflagration revisited. We seek a solution to (45) subject to the boundary conditions

$$\begin{aligned}
& x \to -\infty \quad & T \to T_f, \quad & Y \to Y_f, \\
& x \to +\infty \quad & T \to T_b = T_f + Y_f, \quad & Y \to 0.
\end{aligned} \tag{82}$$

Then with the choice $\gamma = 1$, $\beta = 0$, the reaction term is replaced by a δ-function whose strength is given by (81) with T_* replaced by T_b (other choices for γ and β lead to minor differences); this δ-function can be located at $x = 0$.

The equations on each side of the sheet are easily solved, so that we have

(83) $\quad\begin{array}{lll} x \leq 0 & T = T_f + Y_f e^x, & Y = Y_f - Y_f e^{\text{Le } x}, \\ x > 0 & T = T_b, & Y = 0. \end{array}$

The jump in heat flux across the flame sheet has magnitude Y_f so that

(84) $$Y_f = \sqrt{2 \text{ Le } \mathscr{D}} \cdot \frac{T_b^2}{\theta} \exp(-\theta/2T_b).$$

Since $\mathscr{D} = DM^{-2}$, this formula determines M, the mass flux through the flame. In this way the flame speed is determined, the essential goal of the analysis. The distinguished limit (77) is, in this case, a recognition of the functional dependence of M on the activation energy.

Applications. The description of the deflagration wave is the starting point for a large number of important combustion problems, some of which will now be described. Undoubtedly others would pick different examples: those presented here were chosen on the basis of their importance, their intrinsic interest, their relation to experimental results, and personal bias.

(a) *Stability of plane deflagration.* The problem of the linear stability of the plane deflagration wave is one that is easily formulated but not easily solved. The undisturbed solution is described by simple formulas, and the perturbation equations are linear, but they have nonconstant coefficients. Only in the case of plane disturbances can this difficulty be circumvented in a natural way. The constant density model avoids the problem by decoupling the fluid mechanics from the thermochemistry, adopting as the fundamental equations the set

(85) $\quad \dfrac{\partial T}{\partial t} + \mathbf{v} \cdot \nabla T = \nabla^2 T + \Omega, \qquad \dfrac{\partial Y}{\partial t} + \mathbf{v} \cdot \nabla Y = \dfrac{1}{\text{Le}} \nabla^2 Y - \Omega,$

where $\mathbf{v}$ is a *specified* velocity field. In the case of a deflagration propagating into a quiescent gas, $\mathbf{v}$ can be equated to zero everywhere. The solution for plane deflagration is unchanged by this approximation, but perturbations to it are governed by constant-coefficient equations. These can be easily solved to describe the evolution of disturbances proportional to $e^{iky + \alpha t}$, where y is measured in the plane of the undisturbed flame sheet. T_* is not constant and not equal to the adiabatic flame temperature; it differs from it by a small perturbation.

The key parameter is Le, the Lewis number. Indeed, we have here an example of a Turing system in which instabilities arise because of the interaction of variables that diffuse at different rates. The stability boundaries are shown in the wave-number, Lewis-number plane in Fig. 7 for the case $\theta \to \infty$, $\theta(\text{Le} - 1) = O(1)$. At any wave number there is a band of stability which always includes Le = 1.

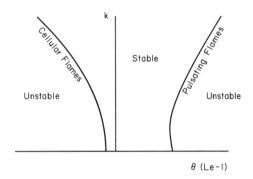

FIG. 7. *Stability regions for plane deflagration.*

The left stability boundary, for which $\alpha = 0$, is associated with cellular flames, a common laboratory phenomenon. The Lewis number is properly defined by the diffusion coefficient of the deficient component of the mixture, and the conduction coefficient of the entire gas. For light fuels such as methane, fuel-lean mixtures can have values of Le small enough to trigger cellular instabilities; for heavy fuels, the mixture must be rich.

Because there is no single "most unstable" disturbance, the manifestation of the instability is an unsteady one, despite the fact that α vanishes on the stability boundary. An equation that describes weakly nonlinear disturbances has been derived and discussed by Sivashinsky; he and his coworkers have also considered several important extensions beyond the problem of plane deflagration [10].

On the right stability boundary $\text{Im}(\alpha)$ does not vanish, so that this is associated with pulsations or travelling waves. Only rather exotic gases, such as lean bromine/hydrogen mixtures, have values of Le large enough to reach this boundary. Nonlinear treatments (Hopf bifurcation) for this and related problems (see (b), below) have been extensively discussed in recent years by Matkowsky and his coworkers, mostly in the SIAM Journal on Applied Mathematics, e.g. [17].

The full problem of flame stability, accounting for the fluid mechanics, remains to be solved. For very small wave numbers the hydrodynamical model of §8 is appropriate, and predicts unconditional instability. If gravity effects are added, long waves can be stabilized, but not short ones. This suggests that stable plane flames can exist with gravity stabilizing the long waves, conduction-diffusion effects (as represented by Fig. 7) stabilizing the short ones. Whether this is true or not, and what shape the stability boundary will take, are questions that could be answered numerically, but have not been. Clavin and his coworkers, e.g. [18], have explored the matter analytically, using small wave-number expansions.

(b) *Unstable fronts in thermites.* Thermites are solids that burn to form solids. They are more than a laboratory curiosity—they promise to be important in the manufacture of exotic materials.

Flame propagation in these substances is governed by the equations that we have written down, but the only permitted velocity fields are uniform flows and the Lewis number is infinite since mass diffusion is insignificant. Because of this, the δ-function strength defined by (81) is not appropriate; however, a modified formula is easily derived, having the same exponential dependence on T_*.

Plane deflagrations are characterized by a jump in Y and a jump in the gradient of T across the flame sheet. They are unstable if θ exceeds some critical value[9], consistent with the fact that large θ gaseous deflagrations are unstable for large Le. Indeed, the stability boundary is like the right boundary of Fig. 7 if the abscissa variable is replaced by θ.

Pulsating fronts are observed in experiments. As a consequence, bands appear in the processed material, presumably associated with incomplete combustion. This is not a feature of a theory in which Y vanishes behind the front, so there is room for a more sophisticated model.

(c) *Hydrogen flame bubbles.* A lean hydrogen/air mixture has a Lewis number significantly less than one so that any flame it supports is highly susceptible to the cellular instability. If the mixture is very weak, the instability is so strong the cells separate from each other, and the flame is fragmented. The fragments consist of small balls of burning gas, or flame bubbles. These are not transient in nature but have permanent form, being sustained by diffusion of hydrogen from the surrounding mixture [19].

The stationary flame bubble can be modelled by

$$(86) \qquad 0 = \frac{1}{r^2}\frac{d}{dr}\left(r^2 \frac{dT}{dt}\right) + \Omega, \qquad 0 = \frac{1}{\text{Le}}\frac{1}{r^2}\frac{d}{dr}\left(r^2 \frac{dY}{dr}\right) - \Omega,$$

the spherically symmetric form of (85). With the subscript f now referring to the far field ($r \to \infty$) the solution can be expressed in terms of r_*, the radius of the spherical flame sheet:

$$(87) \qquad \begin{array}{l} \underline{r > r_*} \quad T = T_f + \dfrac{Y_f}{\text{Le}}\dfrac{r_*}{r}, \qquad Y = Y_f\left(1 - \dfrac{r_*}{r}\right), \\[1em] \underline{r < r_*} \quad T = T_b = T_f + \dfrac{Y_f}{\text{Le}}, \qquad Y = 0. \end{array}$$

The δ-function strength (81) then determines r_*. Note that since Le is small, the flame temperature is much higher than the adiabatic flame temperature.

---

[9] A result inaccessible to large θ asymptotics.

The solution described here is unstable to one-dimensional (radial) perturbations. It is possible that the flow field associated with the buoyancy-induced rise speed characteristic of all observed flame bubbles has a stabilizing effect, but this is an unresolved question.

(d) *Plane burner flames and their stability.* The plane flame discussed at the beginning of this section is unbounded, the fresh mixture being supplied at $x = -\infty$. Plane flames can be generated in the laboratory that are attached to (stabilized by) a burner, so that the source of mixture is at a finite distance from the reaction zone. Various boundary conditions can be adopted at the burner face, but one choice, corresponding to the porous plug burner, is (in dimensional form)

(88) $$T = T_s, \qquad \rho u Y - \mu_{11} \frac{dY}{dx} = \rho u J;$$

the second of these states that a fraction J of the mass flux at the face consists of fresh mixture.

In contrast to the unbounded flame, the mass flux is usually assigned; the distance between the flame sheet and the burner can then be calculated as a characteristic of the solution. The predicted variations of this quantity with mass flux have been successfully compared with experimental data.

Changes in the stability characteristics with mass flux can also be calculated. These characteristics are qualitatively the same as those for the unburned flame, but the presence of the burner introduces quantitative differences. The burner acts as a heat sink, drawing heat from the burning gases, and this displaces the right stability boundary to the left, making it accessible to commonplace mixtures [20]. The precise location of the boundary depends upon the mass flux. As this is decreased, the boundary first moves to the left and then moves back to the right. It is therefore possible to have a stable flame when the mass flux is high, an unstable (pulsating) flame at reduced mass fluxes, and once again a stable flame at yet smaller mass fluxes [21]. Ths behavior has been observed in experiments carried out by Gorman [22].

(e) *Flame tips.* Flame tips can be described on the hydrodynamical scale (§8) and also on the much smaller conduction-diffusion scale. In the context of the constant density model the steady form of (85) may be adopted with $\mathbf{v} = U\mathbf{i}$ (Fig. 8). If $U \gg 1$ so that the gas speed is much bigger than the adiabatic flame speed, the elliptic problem can be reduced to one of parabolic type, with x being the timelike variable. With this kind of simplification the coupled variable density problem is only slightly more difficult, and can be more reasonably compared with experiment. In the *immediate* neighborhood of the tip (the point P in Fig. 8) a local similarity solution is possible, and this leads to a prediction of the tip curvature. Comparisons between this prediction and experiment are shown in Fig. 9.

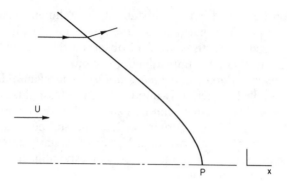

FIG. 8. *Unbounded flame tip.*

(f) *Flames subject to simple strain.* The flow field in which a flame is imbedded is seldom uniform and this can have a powerful influence on the speed and structure of the flame, destroying it in some cases. A straining flow generated by colliding jets is one of the simplest examples (Fig. 10). If the jets are identical, the combustion field is symmetric and there will be two flame sheets or none. If only one of the jets is combustible there can be, at most, one flame sheet.

We assume a flat flame ($\nabla^2 \equiv d^2/dx^2$) and, for the constant density model, use (85) with

(89) $$\mathbf{v} = \beta(-x, y)$$

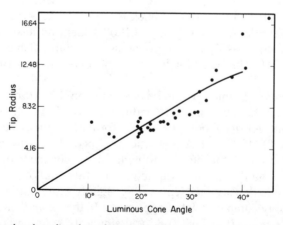

FIG. 9. *Measured and predicted nondimensional flame-tip radius for an axisymmetric flame, as a function of flame cone angle (from T. C. Wagner and C. R. Ferguson, "Bunsen flame hydrodynamics", Combustion and Flame (to appear)).*

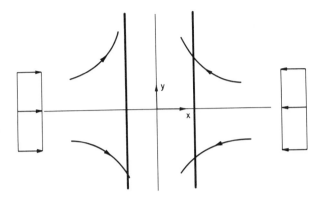

FIG. 10. *Flame(s) in a straining flow generated by colliding jets.*

where β is the straining rate. T and Y are specified in the far field ($|x| \to \infty$).

The distance between the flame sheet and the stagnation point is called the stand-off distance, and its variations with β can be calculated; for the symmetric case the response is shown in Fig. 11. A solution only exists if β is not too large and the nature of the solution at the maximum straining rate depends on the Lewis number. If Le is less than some critical value the stand-off distance (and flame speed) is zero at extinction; for large Le, both are positive. This dichotomy has been confirmed experimentally [23].

For the nonsymmetric case the flame sheet can cross the y-axis as the straining rate is increased, so that the flame speed becomes negative; the mass-averaged velocity at the flame sheet is directed from the burnt gas

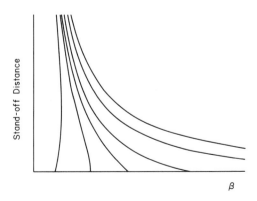

FIG. 11. *Variations of the stand-off distance with straining rate for different values of Lewis number.*

towards the fresh mixture. As in the case of the hydrogen flame bubble, mixture reaches the reaction zone by diffusion.

Several workers, including this author, have hypothesized that negative flame speeds are not achieved in practice; that for reasons not yet understood (instability, perhaps) the flame is extinguished. This provides an explanation of open flame tips, flame quenching near a cold wall, and similar phenomena, and is consistent with the lack of experimental confirmation of negative flame speeds. Recently, negative flame speeds have been observed for *heated* flames [24], but the question is still an open one in so far as adiabatic or nonheated flames are concerned.

Straining flows are an important ingredient of the upward propagation of flames in a tube. A standard measure of whether or not a mixture is flammable is to place it in a vertical tube of circular cross section (diameter 5 cm), and ignite it at the bottom. If flame propagation up the tube is self-sustaining the mixture is said to be flammable, otherwise not. For weak mixtures, close to the flammability limit, the flow generated by the buoyancy of the hot gas behind the flame is a significant ingredient of the total flow field; the hot gas tends to rise like an immiscible bubble, displacing the cold gas ahead of it and generating a straining flow near the center-line. Straining flows are also an important ingredient of turbulence. In view of their importance their effect should be studied using more sophisticated models. Libby and Williams, in a series of papers, e.g. [25], have properly incorporated the fluid mechanics into the description; the next stage is to adopt kinetic models which will permit detailed quantitative comparisons with experiments.

The examples (a)–(e) make it clear that the simple flame-sheet model, although it arises from a very crude approximation of the chemistry, can describe subtle complex behavior that has close parallels in the physical world. This is true even when further abuse is done to the physics by adopting a constant density model—what appears to be abuse is simply the excision of extraneous complexity.

Modelling is a controversial topic since tolerance for deviations from an "exact" formulation varies widely. There are those who, perfectly content to replace twenty reactions of a like number of reactants by a one-step irreversible reaction for one reactant, are hostile to the constant density model, levelling the charge "unphysical". Others, for whom replacing six-fold density variations by zero raises no qualms, criticize the work of Stewart on fast deflagrations (e.g. [26]), because he constructs solutions that are valid when a certain parameter is large, which, in the physical world, has values in the neighborhood of 1.

Science is best served in evaluating a model (particularly one whose ambitions are only qualitative) by considering what it retains (rather than what it discards) and what it is to be used for. The simple chemical model is successful to the extent that the essential effect of the "real" reactions is to generate heat and consume fuel/oxidizer—these ingredients are preserved in the model. If the goal is to understand the generation of oxides of nitrogen in an engine, it will not do. The constant density model retains chemical reaction and differential diffusion and so describes phenomena, such as the cellular instability, which owe their existence to these ingredients. In the context of hydrodynamical interactions, such as those that give rise to the Darrieus–Landau instability, it is unphysical. Stewart's model gives rise to unphysical flame structure, but retains so much of the physics of high Mach number combustion waves that the notion that the dynamical behavior of the solutions (the focus of Stewart's work) can cast no useful light on the dynamics of physical flames, is implausible on its face, and refuted by existing results.

11. Rocket motors, fizz burning, and the right stability boundary. A rocket motor consists of a chamber, in which gases are burnt, attached to a Laval nozzle. In solid rockets the gases are generated by pyrolysis of the propellant and, for double-base propellants, they constitute a homogeneous mixture which supports a deflagration wave (§4); for composite propellants the combustion field is more complicated.

A mathematical description consists of three components: heat conduction within the solid; gas-phase combustion; and the chamber conditions, especially the pressure. The first and last are not part of the classical treatment of deflagration waves and their inclusion can lead to instabilities quite distinct from those discussed in §10. Amongst these are acoustic instabilities in which sound waves bouncing around the chamber are amplified upon reflection from the combustion zone; low frequency (nonacoustic) oscillations for which the corresponding acoustic wavelength is much larger than the chamber, so that conditions there are spatially uniform; and chuffing, a low frequency phenomenon characterized by short firing periods separated by moments of extinction. The first two are reasonably well understood, although the nonlinear manifestation of the second is an area deserving of further study using bifurcation theory and other perturbation methods. Chuffing is understood less well; the popular idea that this is the nonlinear manifestation of low frequency oscillations is yet to be demonstrated.

A type of instability intrinsic to the propellant itself appears to be closely related to the right stability boundary of Fig. 7 and the thermite instability identified in §10. It is most easily demonstrated (mathematically) in the context of fizz burning (sometimes called Zeldovich flameless combustion) in which pyrolysis occurs, but the gases do not burn. Suppose we have a half-space of propellant contiguous with a half-space of its product gases. In a coordinate frame attached to the surface ($x = 0$), one-dimensional equations are:

(90)
$$x \leq 0 \text{ (solid)} \quad \rho_s C_p \frac{\partial T}{\partial t} + C_p M \frac{\partial T}{\partial x} = \lambda_s \frac{\partial^2 T}{\partial x^2},$$

$$x \geq 0 \text{ (gas)} \quad \rho_s C_p \frac{\partial T}{\partial t} + C_p \rho_g v_g \frac{\partial T}{\partial x} = \lambda_g \frac{\partial^2 T}{\partial x^2},$$

$$\frac{\partial \rho_g}{\partial t} + \frac{\partial}{\partial x}(\rho_g v_g) = 0, \quad \rho_g T_g = \text{constant}.$$

$M(t)$ is the mass flux and we have taken C_p to be the same in both phases.
Boundary conditions at the surface are

(91)
$$x = 0 \quad M = \rho_g v_g = f(T) \quad \text{(pyrolysis law)},$$

$$\lambda_s \frac{\partial T}{\partial x}(x - 0) - \lambda_g \frac{\partial T}{\partial x}(x + 0) = MQ;$$

we are concerned with $Q > 0$ corresponding to exothermic pyrolysis.
In the remote portion of the solid,

(92)
$$T \to T_f \quad \text{as } x \to -\infty.$$

The steady solution of (90)–(91), identified by the zero subscript, is

(93)
$$x \leq 0 \quad T_0 = T_f + (T_{*0} - T_f) \exp\left(C_p \frac{M_0}{\lambda_s} x\right),$$

$$x \geq 0 \quad T_0 = T_{*0},$$

$$M_0 = f(T_{*0}), \quad T_{*0} = T_f + Q/C_p.$$

In order to discuss the stability of this solution, it is helpful to write the equations in nondimensional form: with $\lambda_s/C_p M_0$, $\lambda_g/C_p M_0$ the characteristic lengths in the solid and gas, $\rho_s \lambda_s/C_p M_0^2$ the characteristic time, T_f the characteristic temperature, M_0 the characteristic mass flux, $C_p T_f$ the characteristic energy density, we have

(94)
$$x \leq 0 \quad \frac{\partial T}{\partial t} + M \frac{\partial T}{\partial x} = \frac{\partial^2 T}{\partial x^2},$$

$$x \geq 0 \quad \frac{\lambda_g \rho_g}{\lambda_s \rho_s} \frac{\partial T}{\partial t} + \rho_g v_g \frac{\partial T}{\partial x} = \frac{\partial^2 T}{\partial x^2}.$$

In the realistic limit $\lambda_g \rho_g / \lambda_s \rho_s \to 0$ the gas phase description becomes quasi-steady and then (94b) has solution

(95) $$x > 0 \quad T = T_*(t).$$

It is then only necessary to solve (94a) subject to the boundary conditions,

(96)
$$x \to -\infty \quad T \to 1,$$
$$x = 0 \quad M = f(T), \quad \frac{\partial T}{\partial x} = MQ.$$

An examination of infinitesimal perturbations to the steady solution (93), proportional to $e^{\alpha t}$, leads to the result

(97) $$\alpha = \frac{1 - 3p}{2p^2} \pm \frac{1}{2p^2}[(1 - 3p)^2 - 4p^3]^{1/2},$$

where

$$p^{-1} \equiv Qf'(T_{*0}).$$

We have instability if $p < \frac{1}{3}$, i.e. if there is sufficient heat release and/or the pyrolysis law is sufficiently temperature sensitive. At the critical value of p, $\alpha = \pm i\sqrt{3}$, so that the neutral mode is pulsating. This instability, the pulsating instabilities described in §§10a,b, and the instability identified by Rosales and Majda [27] in their model for weak detonation waves[10], all appear to have a common origin.

Highly nonlinear intrinsic instability has been reported in [28] for a cordite propellant. Numerical solutions displaying the nonlinear consequences of the instability can be found in [29].

12. Diffusion flames—the burning fuel drop. Our discussion has so far been restricted to flames supported by homogeneous mixtures. In diffusion flames the reactants (fuel and oxidizers) are supplied separately and mix (by diffusion) and burn simultaneously. Activation energy asymptotics is the only available tool that permits analysis of a wide class of diffusion-flame problems and, unlike premixed combustion, no simple δ-function model can be substituted. The asymptotic structure must be considered in all its (sometimes messy) detail. Most of these details were first developed by Liñán [30]; many of them are described in [5].

A simple example of a diffusion flame arises in the combustion of a liquid fuel drop (Fig. 12). Fuel is generated by evaporation at the drop surface,

---

[10] Their equations are $T_t + C(T^2)_x - q_0 z = \beta T_{xx}$, $z_x = K\Phi(T)z$, where z represents the reactant.

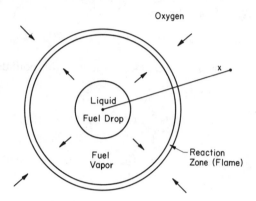

Fig. 12. *Burning fuel drop.*

oxygen is supplied at infinity, and the two come together and burn at a thin spherical surface ($x = x_*$). The governing equations in this case have the nondimensional form

(98)
$$\frac{\partial \rho}{\partial t} + \frac{1}{x^2}\frac{\partial}{\partial x}(\rho u\, x^2) = 0, \qquad \rho T = T_\infty,$$

$$\rho \frac{\partial T}{\partial t} = \Pi T + \Omega,$$

$$\rho \frac{\partial}{\partial t}(X, Y) = \Pi(X, Y) - \tfrac{1}{2}\Omega(1, 1),$$

$$\Pi \equiv \frac{1}{x^2}\frac{\partial}{\partial x}\left(x^2 \frac{\partial}{\partial x}\right) - \rho u \frac{\partial}{\partial x},$$

$$\Omega = \mathcal{D} X Y e^{-\theta/T},$$

$\underline{x = 1}$ $\quad T = T_s,\quad \dfrac{\partial T}{\partial x} = \rho u\, L,\quad \rho u\, X - \dfrac{\partial X}{\partial x} = 0,\quad \rho u\, Y - \dfrac{\partial Y}{\partial x} = \rho u,$

$\underline{x \to \infty}$ $\quad T \to T_\infty,\; X \to X_\infty,\; Y \to 0.$

Here X is the mass fraction of oxidizer, Y the mass fraction of fuel, $\mathcal{D}$ is an assigned parameter called the Damköhler number which depends on the pressure and drop size amongst other things, and L is the latent heat.

When the combustion is steady, mass conservation implies that

(99) $$\rho u\, x^2 = M \qquad \text{(constant)};$$

M is called the burning rate and its variations with $\mathcal{D}$ can be determined by asymptotic analysis. An essential feature of this analysis is the integration (necessarily by numerical means) of flame-sheet equations that are generalizations of (76); the generalization arises from the fact that there are two factors like $[T - T^* + a(x - x_*) + b/\theta]$ multiplying the reaction term, rather than the single factor $(T - T_*)$. The variations of M determined in this way define an S-shaped curve (Fig. 13). Most of the middle branch is known to be unstable—probably it all is.

In the limit $\mathcal{D} \to \infty$ (the Burke–Schumann or equilibrium limit) the solution is particularly simple for then

(100) $$XY = 0$$

to leading order. Between the flame sheet and the drop surface there is no oxidizer ($X = 0$), and between the flame sheet and infinity no fuel ($Y = 0$). To leading order in θ this is true for all solution points on the upper branch of Fig. 13 and this description can be found without analysis of the flame sheet structure (extinction at the top bend of the S-response is only identified when $O(\theta^{-1})$ perturbations are examined). The solution for T is

(101) $$\begin{aligned} \underline{1 < x < x_*} \quad T &= (T_s - L) \\ &+ (T_\infty + 2X_\infty - T_s + L) \\ &\times [(T_\infty + 2X_\infty - T_s + L)L^{-1}]^{-1/x}, \\ \underline{x_* < x < \infty} \quad T &= (T_s - L + 2) \\ &+ (T_\infty - T_s + L - 2) \\ &\times [(T_\infty + 2X_\infty - T_s + L)L^{-1}]^{-1/x}, \end{aligned}$$

where

(102) $$x_* = \ln[(T_\infty + 2X_\infty - T_s + L)L^{-1}]/\ln(1 + X_\infty)$$

defines the flame-sheet location. The burning rate is

(103) $$M = \ln[(T_\infty + 2X_\infty - T_s + L)L^{-1}].$$

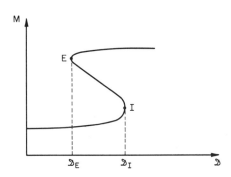

FIG. 13. *Burning response for a fuel drop.*

The points I and E in Fig. 13 are ignition and extinction points. From the lower branch an increase in $\mathcal{D}$ above the value $\mathcal{D}_I$ necessarily causes a jump to the upper branch, and from there a decrease in $\mathcal{D}$ below $\mathcal{D}_E$ causes a jump to the lower branch.

Ignition is characterized by the kind of runaway identified in §6 (see (53)) followed by large temperature changes that occur on an extremely small time scale. Although the ignition of a fuel drop has not been analyzed, similar problems have been studied in a fluid-mechanics-free context. Most present efforts in ignition are concerned with the fluid mechanics, particularly the compression waves that are generated, and the question of whether or not they can lead to detonation.

Mathematically, the extinction problem is not as interesting as the ignition problem, but it is of physical interest and reveals some of the idiosyncrasies of the small Mach number equations. We shall finish this section with a brief discussion of it.

The equation of state (98b) relates the density and temperature and so can be used to eliminate the time derivatives between equations (98a,c). Thus

$$(104) \qquad T_\infty \frac{\partial}{\partial x}(ux^2) = \frac{\partial}{\partial x}\left(x^2 \frac{\partial T}{\partial x}\right) + \Omega x^2,$$

which can be integrated once to yield

$$(105) \qquad T_\infty u x^2 = x^2 \frac{\partial T}{\partial x} + \int_1^x \Omega x^2 \, dx + f(t).$$

This relates the velocity to the other variables; in particular, initial values of u cannot be chosen arbitrarily once initial values of T, X and Y have been assigned. In any problem with initial data not compatible with (105) there will be a rapid adjustment on an acoustic time scale (initial boundary layer). It does not appear that this transient has been discussed.

Suppose that the initial data for T, ρ, X and Y correspond to some point on the steady response curve near E, but $\mathcal{D}$ has been reduced to a value less than $\mathcal{D}_E$. With Ω reduced in this way the initial velocity, described by [105], will be different from the steady-state velocity. The initial response of the combustion field to this imbalance occurs rapidly, being described in terms of the fast time

$$(106) \qquad \tau = \theta^2 t.$$

The significance of this scale is that time derivatives must then be retained in the flame-sheet description ($\partial T/\partial t$ and $\partial^2 T/\partial x^2$ are both $O(\theta)$ within the reaction zone); it is the proper time scale to use in showing that the middle branch of the response is unstable [16].

We write expansions in the form

(107)
$$(T,\rho,u,X,Y) = (T_0,\rho_0,u_0,X_0,Y_0) + \frac{1}{\theta}(T_1,\rho_1,u_1,X_1,Y_1) + \cdots,$$
$$p = \theta^{-2}p_{-2} + \theta p_{-1} + p_0 + \cdots,$$

so that, outside of the reaction zone,

(108)
$$\frac{\partial}{\partial \tau}(T_0, \rho_0, X_0, Y_0) = \frac{\partial}{\partial t}(T_1, \rho_1, X_1, Y_1) = 0,$$
$$\rho_0 \frac{\partial u_0}{\partial \tau} = -\frac{\partial p_{-2}}{\partial x}$$

the latter coming from the momentum equation. Only the velocity and the pressure change significantly; in particular, T_0 is given by the formulas (101).

Since T_0 and ρ_0 do not change outside of the flame sheet, it follows from the surface boundary conditions in (98) that $u_0(x = 1)$ does not change and then (105) implies that

(109) $$u_0 x^2 T_\infty = x^2 \left(\frac{\partial T}{\partial x}\right)_0 + \left[\int_1^x x^2 \Omega \, dx\right]_0 + u_0(x = 1)T_\infty(1 - L/T_s),$$

a result that is valid everywhere. It is the reaction term on the right side of this expression that changes with time and so leads to changes in u_0. More precisely, the reduced value of $\mathcal{D}$ causes the temperature in the reaction zone to drop. This drop is $O(\theta^{-1})$ and causes the exponential factor in the reaction term (cf. (78)) to vanish as $\tau \to \infty$. In this limit u_0 is given by

(110) $$u_0 x^2 T_\infty = x^2 \left(\frac{\partial T}{\partial x}\right)_0 + u_0(x = 1)T_\infty(1 - L/T_s)$$

with T_0, ρ_0, X_0 and Y_0 still unchanged from their initial values. There are large changes in the heat flux $(\partial T/\partial x)_0$ across the flame sheet and, accordingly, large changes in u_0. These can be calculated and Fig. 14 shows the steady-state velocity (before $\mathcal{D}$ is reduced) and the velocity in the limit $\tau \to \infty$ (after $\mathcal{D}$ is reduced).

The physical reasons for the large induced velocity field are clear. The rapid drop in temperature within the reaction zone causes a rapid rise in density $(\partial \rho/\partial t = O(\theta))$. Since the flame-sheet thickness is $O(1/\theta)$ the rate of increase of mass within the sheet is $O(1)$ and the altered velocity field provides the necessary flux; it is driven by the pressure gradient (108c). Once reaction has ceased the collapse of the combustion field continues on the t-scale.

The unsteady processes of ignition or quenching are of fundamental im-

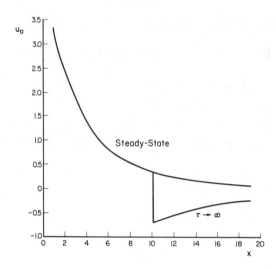

FIG. 14. *Flow speeds for a burning fuel drop in the steady state and the limit* $\tau \to \infty$ ($T_\infty = .07$, $T_s = .06$, $L = .02$, $X_\infty = .47$).

portance in combustion, and our understanding of these processes is far from complete.

13. Diffusion flames—buoyancy effects. In certain diffusion flames, buoyant convection plays an important role in sustaining the flow of oxidizer towards the flame. This is the case for candles and matches. Even when there is forced convection, as in tube-burner flames, buoyant convection can be important, particularly if the flame is large.

Large candle flames ($\gtrsim 4$ cm in height) flicker (the tip oscillates up and down) at a frequency of about 12 Hz, an occasional dinner-table phenomenon. Oscillations of this frequency are seen in tube-burner flames (Bunsen burner flames with the air port closed) and they are a significant component of the energy spectrum of turbulent diffusion flames. These oscillations are probably a manifestation of hydrodynamic instability.

In this section we shall describe an idealized buoyancy-driven flame to give some insight into the flow fields that can arise.

The configuration is shown in Fig. 15. Half-spaces of fuel and oxidizer are separated by a semi-infinite vertical plate for negative x, a diffusion flame for positive x. The flame is sustained by a buoyancy-induced flow field. This model configuration is motivated by tube-burner flames [31].

Provided x is large enough, a boundary-layer (parabolic) formulation is appropriate. If $\mathscr{D}$ is large (cf. §12), there is a flame sheet separating fuel from oxidizer and we may consider a symmetric situation in which this sheet is

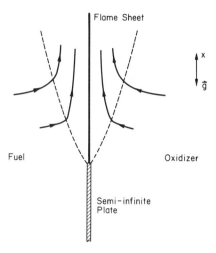

FIG. 15. *Symmetric buoyancy-driven flow.*

coincident with the x-axis. It turns out that the temperature of the sheet is constant, so that rather than determine it by solving the equations for X and Y, we shall assign to it the value T_*. It is then only necessary to solve the equations

(111)
$$\rho \left(u \frac{\partial u}{\partial x} + v \frac{\partial u}{\partial y} \right) = g(\rho_f - \rho) + \frac{\partial}{\partial y} \left(\mu \frac{\partial u}{\partial y} \right) \quad (\mu = \tfrac{4}{3}\kappa),$$

$$\frac{\partial}{\partial x} (\rho u) + \frac{\partial}{\partial y} (\rho v) = 0,$$

$$C_p \rho \left(u \frac{\partial T}{\partial x} + v \frac{\partial T}{\partial y} \right) = \frac{\partial}{\partial y} \left(\lambda \frac{\partial T}{\partial y} \right),$$

$$\rho T = \rho_f T_f = \rho_\infty T_\infty.$$

As always, the subscript f refers to the fresh mixture.

The transport coefficients are assumed to be proportional to the temperature, so that

(112)
$$\frac{\mu}{\mu_*} = \frac{\lambda}{\lambda_*} = \frac{T}{T_*};$$

this is not only more realistic than the adoption of constant values, but leads to simpler equations for a parabolic system in which the equation of state is Charles' law.

The equations can be nondimensionalized using the characteristic length

$(\mu_*^2/\rho_*^2 g)^{1/3}$ and the characteristic speed $(g\mu_*/\rho_*)^{1/3}$, and a similarity solution obtained that satisfies the equations

(113)
$$F''' + \tfrac{3}{4}FF'' - \tfrac{1}{2}F'^2 + H = 0,$$
$$\frac{1}{P}H'' + \tfrac{3}{4}FH' = 0 \quad \left(P = \frac{\mu_* C_p}{\lambda_*}\right),$$

$\zeta = 0$, $F = 0$, $F'' = 0$, $H = 1$; $\quad \zeta \to \infty \quad F' \to 0$, $H \to 0$.

ζ is the similarity variable, defined by

(114)
$$\zeta = (\sigma - 1)^{1/4}\bar{x}^{-1/4}\int_0^{\bar{y}}\frac{\rho}{\rho_*}\,d\bar{y}$$

where $\sigma = \rho_f/\rho_*$ is the density ratio, and barred variables are nondimensional. The physical variables are related to F and H by the formulas

(115)
$$u = \sqrt{g(\sigma - 1)x}\,F'(\zeta),$$
$$T/T_* = \sigma^{-1} + (1 - \sigma^{-1})H(\zeta).$$

The maximum velocity occurs on the center-line, and has value

(116)
$$u_{\max} = \sqrt{g(\sigma - 1)x}\,F'(0)$$

where numerical calculations [31] show that $F'(0) = 0.9$ when $P = .75$. When $x = 4$ cm, $\sigma = 7$, this induced velocity is 138 cm/s. A substantial jet-like flow is created (Fig. 16) which is undoubtedly unstable when x is large enough. The critical Reynolds number for this flow is not known.

This simple, idealized problem has been discussed in order to emphasize the important role that fluid mechanics should play in combustion. Many of the recent developments that have so invigorated the subject have, for simplicity's sake, suppressed the effect of the flame on the flow. One of the most challenging problems faced by combustion scientists is the mutual interaction between flame and flow field.

14. Concluding remarks. In this preliminary chapter we have introduced the equations of the subject, and touched lightly upon some specific problems, in an attempt to convey something of its flavor as a branch of applied mathematics, or a superset of theoretical fluid mechanics. Practitioners in these fields have certain tools that they can, and are willing to use, together with a particular aesthetic, and these color the window through which they view the physical world. The picture presented here was taken through this window. But it is important to understand that there are other, less academic, viewpoints. Indeed, the overwhelming majority of the work done in combustion is mission oriented, closely tied to the physical world, and much

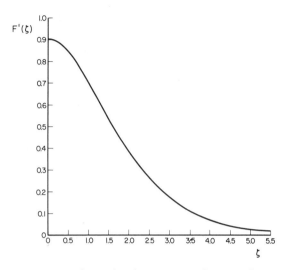

FIG. 16. *Velocity distribution in an infinite candle.*

less elegant. Those who are trying (for example) to understand the three-dimensional potpourri of premixed and diffusion flames which characterize composite solid propellant burning, cannot afford the luxury of elegance in their modelling; the cloistered world of asymptotics, bifurcation, and the like is much too confining.

We theoreticians who wish to make a contribution to combustion are well advised to be aware of this other, bigger world. Otherwise we run the risk of merely providing amusement for the small community of our peers, without making any lasting scientific contribution. The questions that we should be addressing cannot be gleaned from the pages of the SIAM Journal on Applied Mathematics (no matter how fine the papers there might be); the proceedings of the biennial international combustion symposia, organized by the Combustion Institute, are a much more appropriate source. The well motivated and informed applied mathematician can make vital contributions to combustion of course—the elucidation of cellular instabilities (§10) is evidence enough.

Within these pages, in hinting at future direction, we have emphasized kinetic modelling since, quite properly, there is presently much interest in constructing new models. Twenty-five years ago, Lighthill's model of an ideal dissociating gas played a valuable role in problems of aerotherochemistry associated with the early days of space exploration. In the last ten years, large activation energy Arrhenius kinetics has led to significant advances in our understanding of the fundamental behavior of premixed flames and diffusion flames. We can be optimistic that new kinetic models, provided they are well motivated, will add significantly to our understanding of combustion in the next decade.

The chapters that follow this one describe other areas that promise important future developments. As was remarked earlier, sensitivity analysis, the subject of Chapter II, can play an important role in kinetic modelling by identifying those reaction paths that are most important. Turbulence, the subject of Chapter III, is so difficult and so ubiquitous that it will always be important. Here the types of idealization so dear to the applied mathematician's heart, such as flame-sheet models, have already played a useful role.

Wave propagation, fast flames (flames with nonvanishing Mach number) and detonations are of enormous importance, and are discussed in Chapters IV and V. A key problem here is DDT—deflagration to detonation transition. Whether it occurs in coal- or grain-dust laden atmospheres, or in damaged solid rocket propellants, it is often to be avoided, if possible, and this requires an understanding of the phenomenon that is lacking at the present time.[11]

These various discussions identify several areas of importance, but no attempt has been made to identify all of the major topics encompassed by theoretical combustion, or even to list the important unsolved problems in the subject. Instead, we shall finish this chapter by briefly mentioning the single most important unresolved question: roughly speaking, it is "Why does a flame go out?". In the discussion of straining flows in §10, flammability limits were mentioned. These are measured by placing the gas in a standard flammability tube and igniting it at one end (assuming the mixture is strong enough for this to be possible). The flame that is generated will either travel through the tube in a self-sustaining fashion, or it will go out after a short period of propagation. The mixture strength that divides gases with these two behaviors defines the flammability limit. In the case of un-

---

[11] An interesting feature of DDT is the reemergence of the cold-boundary difficulty (mentioned in §4) as an issue. Much of the present work in flame propagation has been able to ignore it—the flame-sheet model of §10 eliminates all reaction at temperatures less than the flame temperature. But the structure of fast flames includes an induction zone characterized by weak reaction at low temperatures, and the difficulty must be addressed. One way is to treat all flames as bounded—in some generalized sense they are thought of as being attached to a burner. Another way is to tinker with the kinetics by introducing an ignition temperature, below which there is no reaction. There is presently a controversy between advocates of these different solutions: the burner school claims that an ignition temperature is not a fundamental property of the gas so that an unknown, unphysical parameter is introduced into the description; the ignition school notes that burner flames require the specification of boundary data which is unknown in, say, the case of a flame traveling down a tube. To this writer's mind the criticism of an ignition temperature is valid (although it does not follow that the predictions of such a model are without physical value); but the notion that all flames are bounded is equally unappealing.

sustained propagation we have an evolving flame that extinguishes, and how and why this occurs is poorly understood.

The same question embraces the quenching of flames by proximity to cold surfaces (for example, at a burner rim), or the quenching that occurs at Bunsen burner flame tips, which allows unburnt gas to pass through a hole at the tip. Suggestive results have been obtained, and important effects identified, but we have yet to achieve a synthesis of kinetic modelling, an understanding of the effects of nonuniform thermal and velocity gradients, and whatever other ingredients that are required, which could allow us to understand and predict these types of phenomena in a general way. Even in the case of fuel-drop "extinction" identified in §12, true extinction is not a feature of the model. Instead, there is merely a transition to a reduced burning rate on the lower branch. A mathematical framework within which we can understand true extinction is elusive.

REFERENCES

[1] JANAF *Thermochemical Tables*, prepared by Dow Chemical Co., U.S. Department of Commerce, National Bureau of Standards, Institute for Applied Technology, Washington, DC, 1966.

[2] J. F. CLARKE AND J. B. MOSS, *On the structure of a spherical H_2-O_2 diffusion flame*, Combustion Sci. Tech., 2 (1970), pp. 115–129.

[3] C. E. WESTBROOK AND F. L. DRYER, *Simplified reaction mechanisms for the oxidation of hydrocarbon fuels in flames*, Combustion Sci. Tech., 27 (1981), pp. 31–43.

[4] N. PETERS (private communication). See also F. L. Dryer and I. Glassman, *High-temperature oxidation of CO and CH_4*, Fourteenth Symposium (International) on Combustion, 1973, The Combustion Institute, Pittsburgh, p. 987.

[5] J. D. BUCKMASTER AND G. S. S. LUDFORD, *Theory of Laminar Flames*, Cambridge Univ. Press, Cambridge, 1982.

[6] J. W. DOLD, *Analysis of the early stage of thermal runaway*, in press.

[7] A. K. KAPILA, *Homogeneous branched-chain explosion: initiation to completion.* J. Engrg. Math., 12 (1978), pp. 221–35.

[8] W. HOCKS, N. PETERS AND G. ADOMEIT, *Flame quenching in front of a cold wall under two-step-kinetics*, Combustion and Flame, 41 (1981), pp. 157–70.

[9] J. D. BUCKMASTER AND G. S. S. LUDFORD, *Lectures on Mathematical Combustion*, CBMS Regional Conference Series in Applied Mathematics, 43, Society for Industrial and Applied Mathematics, Philadelphia, 1983.

[10] G. I. SIVASHINSKY, *Instabilities, pattern formation, and turbulence in flames*, Ann. Rev. Fluid Mech., 15 (1983), pp. 179–99.

[11] M. S. UBEROI, *Flow field of a flame in a channel*, Phys. Fluids, 2 (1959), pp. 72–78.

[12] YA. ZELDOVICH, A. G. ISTRATOV, N. I. KIDIN, AND V. B. LIBROVICH, *Flame propagation in tubes: hydrodynamics and stability*, Combustion Sci. Tech., 24 (1980), pp. 1–13.

[13] J. BUCKMASTER AND A. CROWLEY, *The fluid mechanics of flame tips*, J. Fluid Mech., 31 (1983), pp. 341–361.

[14] A. K. KAPILA, *Asymptotic Treatment of Chemically Reacting Systems*, Pitman, Boston, 1983.

[15] J. F. CLARKE, *Parameter perturbations in flame theory*, Prog. Aerospace Sci., 16 (1975), pp. 3–29.
[16] J. BUCKMASTER, A. NACHMAN, AND S. TALIAFERRO, *The fast-time instability of diffusion flames*, Physica D, 9 (1983), pp. 408–24.
[17] B. J. MATKOWSKY AND D. O. OLAGUNJU, *Spinning waves in gaseous combustion*, SIAM J. Appl. Math., 42 (1982), pp. 1138–56.
[18] P. PELCE AND P. CLAVIN, *Influences of hydrodynamics and diffusion upon the stability limits of laminar premixed flames*, J. Fluid Mech., 124 (1982), pp. 219–37.
[19] B. LEWIS AND G. VON ELBE, *Combustion, Flames and Explosions of Gases*, 2nd edition, Academic Press, New York, 1961.
[20] S. B. MARGOLIS, *Bifurcation phenomena in burner-stabilized premixed flames*, Combustion Sci. Tech., 22 (1980), pp. 143–69.
[21] J. BUCKMASTER, *Stability of the porous plug burner flame*, SIAM J. Appl. Math., 43 (1983), pp. 1335–49.
[22] M. GORMAN (private communication).
[23] M. TSUJI AND I. YAMAOKA, *An experimental study of extinction of near-limit flames in a stagnation flow*, in First International Specialists' Meeting of the Combustion Institute, Section Française du Combustion Institute, 1981, pp. 111–16.
[24] S. H. SOHRAB, C. Y. ZYE, AND C. K. LAW, *An experimental investigation on flame interaction and the existence of negative flame speeds*, Twentieth Symposium (International) on Combustion, 1985, The Combustion Institute, Pittsburgh.
[25] P. A. LIBBY AND F. A. WILLIAMS, *Strained premixed laminar flames with two reaction zones*, Combustion Sci. Tech., 37 (1984), pp. 221–52.
[26] D. S. STEWART AND G. S. S. LUDFORD, *Fast deflagration waves*, J. de Mécanique, Théorie et Appliqué, 3 (1983), pp. 463–87.
[27] R. R. ROSALES AND A. J. MAJDA, *Weakly nonlinear detonation waves*, SIAM J. Appl. Math., 43 (1983), pp. 1086–1118.
[28] J. D. HUFFINGTON, *The unsteady burning of cordite*, Trans. Faraday Society, 50 (1954), pp. 942–52.
[29] D. E. KOOKER AND C. W. NELSON, *Numerical solution of solid propellant transient combustion*, J. Heat Transfer, 1 (1979), pp. 359–64.
[30] A. LIÑÁN, *The asymptotic structure of counterflow diffusion flames for large activation energies*, Acta Astronautica, 1 (1974), pp. 1007–39.
[31] J. BUCKMASTER AND N. PETERS, *The infinite candle—a study in buoyancy-induced flows*, to be published.

CHAPTER II

Sensitivity Analysis of Combustion Systems[1]

HERSCHEL RABITZ

1. Introduction. Mathematical modelling tools are being applied at an increasing rate to the problems arising in the area of combustion. Engineering modelling on full scale combustors is a complex subject typically encompassing chemistry, fluid mechanics (including turbulence effects) and certain aspects of chemical engineering. At the other extreme of basic science, the elementary reactions in combustion systems have been studied at the atomic and molecular level using procedures from quantum mechanics. The gulf between the fundamental atomic scale events and the characteristic behavior of macroscopic combustors is enormous. Therefore, problems in this area are typically broken into smaller segments and studied individually for their own merit and understanding. At this time, the goal of truly integrating together fundamental microscopic processes into an overall numerically implemented mathematical model has not been achieved, although enormous strides are being made towards this direction. The present chapter is concerned with a general question that invariably arises at any level in mathematical modelling of combustion processes as well as in other areas. In this regard, the paper will not focus directly on issues of mathematical modelling per se, but rather on the analysis of such modelling efforts. Examples will be drawn from cases involving transport as well as chemistry, but the illustrations will certainly not sample all aspects of mathematical modelling in the combustion field. For the most part, the treatments will focus on macroscopic issues, although some comments about relevant microscopic atomic level events will also be briefly discussed. It is important to keep in mind that the techniques of sensitivity analysis discussed in this chapter are quite general and they have broad applicability in virtually any area of mathematical modelling.

Realistic models of combustion processes inherently require numerical simulations due to the complicated and generally nonlinear nature of the physical processes. In order to understand the role of sensitivity analysis

---

[1] The author would like to acknowledge support for this work by the Department of Energy, the Office of Naval Research and the Air Force Office of Scientific Research.

associated with such modelling, it is first necessary to appreciate that one may perform modelling without the use of sensitivity analysis but sensitivity analysis may not be done without the performance of modelling. Indeed, all too often modelling is performed without the utilization of sensitivity techniques, and it is the general thesis of this paper that such efforts will invariably fall short of their full utility without the implementation of a sensitivity analysis.

Sensitivity analysis has tended to be associated with an error analysis of the modelling process. Although this is a powerful application of the technique, in a real sense it is only a small part of how sensitivity analysis may be applied. In many cases, the physical understanding gained by performing a sensitivity analysis may outweigh issues of model predictive uncertainty. When actually implementing numerical models, it is quite common to run the program for at least several choices of parameter values characterizing the physical system. In a sense, this type of effort represents a crude form of sensitivity analysis. In general, if one could carry out an arbitrary number of calculations throughout the relevant parameter space, then the assembled results would in principle contain all possible information about the class of systems under study. This *global mapping* issue is notoriously difficult for realistic problems typically involving a high dimensional space of parameters. Sensitivity analysis usually aims at a more modest goal of performing a detailed exploration of the behavior around a local operating point in parameter space. Nevertheless, global issues have a direct relation to sensitivity analysis, and this subject will be brought out in the paper. In previous work, the desirability of performing a formal sensitivity analysis has frequently been recognized, but such calculations were often put aside due to the apparent sheer computational expense involved. At the present time, the numerical techniques for implementing sensitivity analysis are rapidly developing, but it is possible to make a rather broad statement concerning this matter. In particular, if numerical implementation of the mathematical model is feasible, then the accompanying sensitivity analysis is also quite affordable.

The historical roots of sensitivity analysis go back many years and the strongest connection is perhaps best associated with the development of stability analysis in the last century. Stability analysis grew up in the consideration of ordinary differential equations, particularly those arising in classical mechanics. Stability issues still form a central part of sensitivity analysis although the subject now encompasses a much wider perspective. There are available two excellent general texts [1], [2] on the subject of sensitivity analysis and a number of review articles exist in the context of the particular applications involved [3]–[6]. The present paper is in this same vein with the aim of drawing the latest sensitivity techniques to the attention of the combustion community. However, this work does not aim to be a comprehensive review in the combustion field. Specific illustrations and references will be cited as a guide to the reader and as an introduction to the literature on the subject.

Regardless of the physical problem, there remains a degree of commonality when sensitivity analysis is utilized. This viewpoint may be understood by examining Fig. 1 which is a schematic diagram of mathematical modelling with an emphasis on the role of system parameters. Throughout this paper, the parameters entering into a model will be denoted as $\alpha_1, \alpha_2, \cdots$. Frequently these parameters are constants with respect to coordinates and time, although there are cases when it is expedient for the parameters to be treated as distributed variables. This issue will be dealt with under the label of functional sensitivity analysis. In combustion systems, the parameters will typically consist of rate constants, transport coefficients, thermodynamic variables, initial species concentrations and parameters entering into the boundary conditions. It is evident that some of these parameters may actually be controlled in the laboratory, while others are only known through the performance of preceding experiments or calculations. From a physical viewpoint, parameters which may be adjusted in the laboratory naturally lead to the consideration of control theory concepts and stability issues, while the remaining "natural" parameters are usually associated with me-

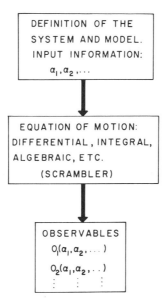

FIG. 1. *A schematic diagram of mathematical modelling from a perspective relevant to sensitivity analysis. The observables O_i could be chemical concentrations, the energy, etc. or functionals of these quantities. From a sensitivity point of view, the model equations act as a scrambler for the detailed parametric information entering the model.*

chanistic and transport details. Regardless of this physical distinction on the meaning of the parameters, their mathematical analysis is basically the same and this point will become clear in the next section where all system sensitivities will be put into a common framework.

Returning again to Fig. 1, it is tacitly assumed that each of the parameters $\alpha_1, \alpha_2, \cdots$ entering into the mathematical model has a clear physical meaning at the input stage. The next step in executing the model consists of solving the appropriate dynamical equations denoted by the second box in Fig. 1. The nature of these equations and their numerical implementation depends on the specific details of the problem, but for our purposes at this juncture, the central point of this step is the scrambling effect that solving the equations has on the former precise knowledge of the roles played by each parameter. In other words, the output depicted in the third box in Fig. 1 is an unknown, and generally highly complicated function (or functional in the case of distributed parameters) of the system parameters. The complexity of this problem is evident even in the simplest case of a linear equation with constant coefficients. Despite the linearity of the problem, the solutions will generally depend nonlinearly on the system parameters. Typically, the equations in combustion are themselves nonlinear, therefore even further obfuscating the role of the parameters. In many respects, this issue is at the heart of mathematical modelling problems since knowledge of how the output observables depend on the system input parameters is ultimately always the physical question of primary concern. These observations provide the basic motivation for introducing sensitivity analysis which in simple terms aims to provide information on the relation between system input parameters and output observables. We will see later that understanding this connection has many subtleties ultimately leading to the conclusion that all the system variables (dependent and independent) are best treated on the same footing.

The observables implied in Fig. 1 may often be the direct solution to the relevant differential equations for the problem. For example in kinetics one can in principle measure chemical concentrations, temperature profiles, velocity fields, etc. In other cases, the actual output of interest may be represented as a function or functional of the raw solutions to the differential equations. For example, this would be the case if bulk thermodynamic properties were the actual system observations in the laboratory. In other cases, a single operating performance index, such as might arise in an engineering application with a combustor, would be the desired output. In the latter situation, the system parameter roles are even further obscured due to the typical added convolution needed to obtain an overall performance index. As a general rule, one might expect that further averaging to get coarser grained observables would naturally lead to reduced system sensitivity with respect to the input parameters. Although this is likely to be the case, there is no way of quantitatively estimating this effect without performing a formal sensitivity analysis. Secondly, there may arise situations when the output is surprisingly sensitive to some particular system parameters despite a high degree of averaging. Indeed, due to the highly nonlinear nature of the parameter interconnections, it may happen that some parameters will have their roles strongly enhanced by the convolution process. In this regard, a fundamental question concerns the degree to which microscopic atomic level information is relevant at the macroscopic level of kinetic/transport phenomena. Surely some atomic level parametric information must remain since different fuels and oxidizers can have widely varying combustion behavior. There is little known about the answer to this question, but sensitivity analysis can in principle provide quantitative insight into the issue.

Without going into the mathematical details at this stage, it is useful to discuss some of the general applications of sensitivity analysis. The first coefficients encountered in actually numerically performing a sensitivity analysis provide a direct measure of the relationship between the input parameters and output observables. These *elementary* coefficients give considerable basic insight into which aspects of the model are important in relationship to particular observations of concern. The latter comment is of critical importance when using general sensitivity information, since certain parameters may be important for a subset of observations while having little significance for others. This issue must always be kept in mind lest false conclusions be drawn regarding model performance. Another application of sensitivity analysis already referred to above involves the quantitative estimation of error propagation from the parameter space to the set of observables. Although it is standard practice to perform such an estimate in reporting experimental data, unfortunately one virtually never sees these estimates appearing along with modelling calculations. Sensitivity analysis provides a rather routine methodology for this latter purpose. At a more advanced level, there are a host of physical questions best addressed by interchanging the role of the system dependent and independent variables. For example, it may be of interest to deduce how a given parameter is sensitive to a potential observation under consideration for measurement in the laboratory. Another case concerns the relationship between two or more system parameters such that they may be altered but still leave a selective set of observations invariant. Both of these situations are especially relevant when considering problems of inversion of laboratory data back to a fundamental set of parameters. Still another type of sensitivity issue concerns overall system dynamical stability. Stability behavior in the framework of sensitivity analysis is probed by a particular set of sensitivity coefficients. As commented earlier, sensitivity analysis has much of its basic historical foundations in the stability area. Despite the historical chronology, one may show that stability behavior and parametric sensitivity analysis are on an equal footing and neither one has a physical or mathematical precedent over the other.

From an experimentalist's perspective, mathematical modelling is most useful when it can directly address characteristic features of data evident in the laboratory. For this purpose, a feature or objective function sensitivity analysis may be performed to relate parameters that are meaningful in the laboratory to those naturally entering into the system model. Another sensitivity related issue arises when it is recognized that the normally deterministic dynamical equations in fact should have some degree of stochastic character. Fluctuations are always present in real systems and important sensitivity issues arise due to the presence of fluctuations, particularly in reactive problems capable of runaway or explosive behavior. Much of the traditional machinery of sensitivity analysis may be carried over to stochastic differential equation models, and some initial work has been carried out towards this goal. Another question that often arises when utilizing sensitivity analysis concerns the parameters that are actually *left out* of the model. Any model, no matter how sophisticated, will inherently be missing some degree of structure. Therefore, the matter of sensitivity to missing model components is a natural topic. Although ultimately one must actually add the new components to truly test their effects, some valuable guidance in this regard can be obtained from appropriate sensitivity coefficients. From exactly the opposite perspective, considerable interest has been expressed in "lumping" complex models in order to obtain reduced pictures containing fewer parameters and dependent variables. Although lumping will of necessity be an approximation, it can provide an extremely valuable tool for the practical treatment of real engineering level combustion problems. Lumping of complex models still remains very much an art, but sensitivity information can be a helpful guide in this area. Finally, sensitivity analysis has a role to play when considering the numerical execution of the model. In particular, one of the most serious questions in the numerical implementation of any model concerns the choice of approximations associated with spatial and temporal discretization of the equations. From a sensitivity point of view, this problem is addressed by seeking the sensitivity of the output quantities to the location or spacing of the mesh points.

The narrative comments above will be amplified on in the following sections, and it is important to keep in mind that the list of applications of sensitivity techniques will likely continue to grow as they become of wider use. The remainder of this paper is organized along topical lines in §§2 and 3. Examples will be drawn from temporal, stationary and space-time problems. Finally, in §3, some broader issues primarily of future concern to sensitivity analysis applications in combustion will be discussed.

2. Sensitivity techniques.
2.1. Model equations.
The problems discussed later in this section will not deal with turbulence phenomena or complex geometrical factors associated with real combustion systems. Modelling of these latter problems is still a subject of intense development and it is anticipated that sensitivity techniques will again be useful as the modelling tools become more well developed. The general class of problems treated below may be described by the following reaction-diffusion equation:

$$(1) \quad \frac{\partial c_i}{\partial t} = -v \frac{\partial c_i}{\partial x} + \frac{\partial}{\partial x} D_i \frac{\partial}{\partial x} c_i + R_i(\mathbf{c}, \mathbf{k}), \quad i = 1, \cdots, N$$

where for simplicity one spatial dimension is chosen for illustration. In this equation, $\mathbf{c}$ is the vector of chemical species concentrations, v is the flow velocity, D_i is the ith diffusion coefficient, R_i is the chemical rate term which is generally nonlinear in the concentrations and depends on the vector of rate constants $\mathbf{k}$. The system of equations (1) is supplemented by a set of initial conditions and boundary conditions

$$(2a) \quad c_i(x, 0) = c_i^0(x),$$

$$(2b) \quad \mathcal{B}_i\left(\mathbf{c}, \frac{\partial \mathbf{c}}{\partial x}\right)\bigg|_{x_L, x_R} = 0, \quad x_L \leq x \leq x_R.$$

The boundary condition in (2b) could take on a variety of forms including nonlinearities. Equation (1) may also be augmented by a similar energy balance equation which is usually expressed in terms of the temperature being the dependent variable. In addition where appropriate an equation for conservation of momentum may also be prescribed. Although the treatment in this paper will be confined mainly to systems described by (1), it is worth mentioning that sensitivity analysis may be put on a more abstract and general level by specifying the equations of motion to have an operator form $\mathbf{L}(\mathbf{c}, \boldsymbol{\alpha}) = 0$ where the vector operator $\mathbf{L}$ is typically nonlinear. The mathematical logic in the latter case is identical to the treatment described below, thereby indicating how sensitivity analysis may be applied to a broad class of problems. Thus far, the system in (1) has been assumed to be deterministic with all of the parameters and dependent variables having no stochastic character. In §2.4 this restriction will be relaxed with a stochastic extension of (1).

Reaction-diffusion models defined by the system in (1) will be referred to below as the space-time problem. Two natural restricted cases will also be discussed: (i) the temporal case without spatial diffusion or convection and (ii) the steady state limit without any time dependence. It has been quite popular in the past to treat purely temporal cases containing extensive amounts of chemistry and those involving transport with little chemistry. Although the historical reasons for this disparity are undoubtedly connected with the numerical difficulties of solving the full equations, it nevertheless must be recognized that real problems will in general be characterized by complex interactions of chemistry and transport processes. The parameters in these problems may arise in the differential equations, boundary conditions and/or initial conditions. The parameter vector $\boldsymbol{\alpha}$ is of length M, and it contains all of these possible parameters. In typical problems one usually encounters the situation $M \gg N$, and this condition has serious consequences for the viability of combustion modelling as well as the practical details of performing a sensitivity analysis.

Before performing a sensitivity analysis, it is tacitly assumed that the system in (1) can be numerically solved to some appropriate level of accuracy. The details of these procedures are not the direct topic of this article, but it will become apparent below that this matter has a significant impact on the practical procedures for performing sensitivity analysis. This comment follows since ideally it is desirable to produce algorithms which are optimum both for solving the dynamical equations (1) as well as the appropriate sensitivity equations. At this juncture, it suffices to realize that (1) is typically discretized in the spatial domain by any of a variety of techniques to produce a large set of coupled ordinary differential equations. The resultant system is generally stiff necessitating the use of implicit integration procedures. For problems where only the steady state solution is sought, it is possible to consider various iterative techniques on the steady state equations (or equivalently integrate the temporal equations to steady state).

At this point, we must now define the type of system sensitivity to be discussed below. In principle, a variety of quantities might be taken as a measure of sensitivity. The most useful of these techniques appear to be gradient based methods which assume that the relevant equations and their solutions are differentiable with respect to the system parameters. The only anticipated difficulty with this perspective would occur at bifurcation points in parameter space where multiple solutions can exist. We therefore assume that the system in (1) will be solved at an operating point $\boldsymbol{\alpha}^0$ in parameter space where the system is differentiable with respect to parameters. It is these derivatives, and the subsequent calculus associated with them, which we will refer to as sensitivity analysis.

2.2. Elementary sensitivity coefficients. In keeping with the above discussion, we now want to consider the *elementary* sensitivity gradients ($\partial c_i/\partial \alpha_j$), ($\partial^2 c_i/\partial \alpha_j \partial \alpha_l$), etc. These gradients are sometimes just referred to as sensitivity coefficients, but for the purpose of distinguishing them from some other types of sensitivity gradients they will be referred to as elementary.

It is quite easy to see that knowledge of these gradients would give a direct measure of how the ith observable concentration is related to particular parameters in the model. It is also common practice to normalize these gradients in some fashion (e.g., $\partial \ln c_i/\partial \ln \alpha_j$) in order to remove any extraneous biasing due to the numerical values of the concentrations or parameters. Clearly a large number of gradients can arise; in first order there are NM and in second order NM^2, etc. The implied Taylor series generated by the high order coefficients would produce information about successively larger regions of parameter space around the operating point α_0 at which the gradients are evaluated. There is of course no guarantee that such a series would converge and secondly the calculations would be enormously tedious. In addition, if global parameter space mapping issues are of concern, then traditional sensitivity techniques are not ideally suited for the task. In many respects global mapping is an *inherently* difficult problem, but there may be better ways of treating global issues (cf. §3.6). In practice, it has been found that the first order gradients often give quantitatively valid insight into parametric input-output questions. Nevertheless, there are cases where second order derivatives may be required and procedures for calculating them in many cases are available. Furthermore, in chemical kinetics arguments may be put forth to at least establish a qualitative understanding of the second order coefficients from the behavior of the first order ones. This connection is depicted in Fig. 2 which points out that for chemical concentrations in a finite system there will always be a maximum concentration value and there is always the minimum $c_i \geq 0$. Therefore, the first

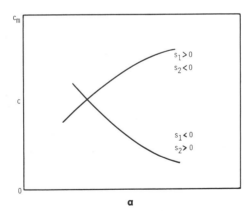

FIG. 2. *Two possible cases of a chemical concentration as a function of an arbitrary model parameter. The symbols s_1 and s_2 denote, respectively, first and second order sensitivity coefficients with respect to the parameter α. The boundedness of the concentration $0 \leq c \leq c_m$ forces a qualitative correlation between the first and second order sensitivity coefficients indicated in the figure.*

and second order derivatives tend to be correlated in the sense of having opposite signs in order to maintain the concentration within its bounds. This qualitative observation may be useful at times, but actual numerical calculations would have to be employed if specific values are needed. The treatment below will be organized according to the type of physical models involved.

(1) *Temporal problems.* The pure temporal limit of (1) has the following form:

$$\frac{\partial c_i}{\partial t} = R_i(\mathbf{c}, \mathbf{k}), \quad i = 1, \cdots, N,$$
(3)
$$c_i(0) = c_i^0,$$

with the parameters consisting of rate constants and initial conditions. The sensitivity equations are obtained by taking the first variation of this system of differential equations to produce

(4a) $$\frac{\partial}{\partial t}\left(\frac{\partial c_i}{\partial \alpha_j}\right) = \left(\frac{\partial R_i}{\partial \alpha_j}\right) + \sum_l \left(\frac{\partial R_i}{\partial c_l}\right)\left(\frac{\partial c_l}{\partial \alpha_j}\right)$$

where

(4b) $$\left(\frac{\partial c_i(0)}{\partial \alpha_j}\right) = \begin{cases} 0 & \text{if } \alpha_j \text{ is a rate constant,} \\ \delta_{ij} & \text{if } \alpha_j = c_j^0. \end{cases}$$

These equations are valid provided the parameters are not dependent on time. The first term on the right-hand side of (4a) is nonzero provided R_i is explicitly dependent on the jth parameter. The last term in (4a) represents the implicit dependence of R_i on α_j through the concentrations. In general, (4) is a coupled set of linear inhomogeneous differential equations with *nonconstant* coefficients. This point follows since the coefficients in (4a) can generally depend explicitly on the solution to (3). Equation (4) may be solved by a number of techniques, and the optimum choice is still a matter for development. This latter comment also applies to the analogous generation of elementary sensitivity coefficients for spatially dependent problems. The most direct approach consists of numerically integrating (4) simultaneously with (3) since the former are coupled to the solutions of (3). Although this method may be simple from a programming point of view, considerable expense can be involved due to the large number of equations present in (4). However, a variation on this approach seems quite attractive by noting that $\partial R_i/\partial c_l = J_{il}$ entering into (4) are elements of the Jacobian matrix for the system in (3). Many of the stable implicit schemes for solving (3) would already calculate the Jacobian and carry forth exactly the necessary numerical manipulations needed in practice to solve (4). Therefore, provided the same mesh points employed in (3) are applicable to (4), then it appears that an efficient solver can be generated [7]. The cautionary provision in the last sentence arises since it is possible for the sensitivity coefficients to vary rapidly or at least behave differently from the chemical concentrations as a function of time. However, in practice the two quantities seem to be coupled in their behavior, making the above technique practical and attractive. There are other ways of treating (4): in particular, the linear nature of the equations can be advantageously utilized. In this regard, an Analytically Integrated Magnus (AIM) method has been developed based on calculating the system Green's function which is formally given by the relation

$$\mathbf{G}(t, t') = T \exp\left[\int_{t'}^{t} \mathbf{J}(\tau)\, d\tau\right] \quad (5)$$

where T is the time ordering operator commonly utilized in quantum mechanics for similar problems. The matrix $\mathbf{G}$ is physically very important and will be discussed in detail in §2.5 below. For our purposes here, it is sufficient to recognize that $\mathbf{G}$ satisfies the equation

$$\left(\mathbf{1}\frac{\partial}{\partial t} - \mathbf{J}\right)\mathbf{G}(t, t') = \mathbf{1}\delta(t - t') \quad (6)$$

and comparison of (4) and (6) shows that the elementary first order sensitivity coefficients may be expressed in terms of the Green's function as

$$\frac{\partial c_i(t)}{\partial \alpha_j} = G_{ij}(t, 0)\delta_{\alpha_j, c_i^0} + \sum_l \int_0^t G_{il}(t, t') \frac{\partial R_l(t')}{\partial \alpha_j}\, dt'. \quad (7)$$

The Kronecker delta-function δ_{α_j, c_i^0} entering into the first term on the right-hand side of (7) is only nonzero when the jth parameter is equal to the ith initial species concentration. For rate constant sensitivity, this latter term is therefore zero.

The sensitivity coefficients can exhibit a wide variety of behavior from problem to problem and even within a given system. Figure 3 shows examples of sensitivity coefficients from the $CO/O_2/H_2O$ oxidation problem [8]. Some reactions may enhance the concentration of a species at one time and diminish it at another time. In addition, if a given elementary step in a reaction

$$\text{reactants} \underset{k_r}{\overset{k_f}{\rightleftarrows}} \text{products}$$

is in dynamic equilibrium during some period of evolution of the system, then all of the chemical species will be just a function of the equilibrium constant $c_i = c_i(k_f/k_r)$. Under these conditions it is a simple matter to verify that the normalized forward and reverse sensitivity coefficients are equal in magnitude but opposite in sign $\partial c_i/\partial \ln k_f = -\partial c_i/\partial \ln k_r$. Such mirror im-

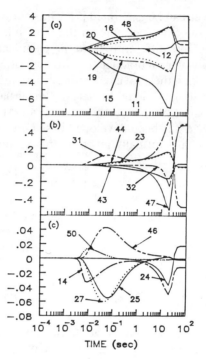

Fig. 3. *Normalized first order elementary sensitivity coefficients $\partial \ln CO/\partial \ln k_j$ for the oxidation system $CO/O_2/H_2O$. The numbers labelling the curves refer to steps in a multireaction scheme [8]. Panels (a), (b) and (c) successively include sensitivities of smaller magnitude. It is apparent in (a) that the forward and reversed reaction constants k_{15} and k_{16} for a given step of the mechanism are in dynamic equilibrium due to the symmetry of the sensitivity curves.*

aging (or the lack of it) in the sensitivity coefficients is easy to identify in their time-dependent graphs and gives a simple characterization of particular elementary steps. It can happen that a given step may be in dynamical equilibrium during some time intervals and no longer show this characteristic behavior at other times.

Although most combustion systems which have been studied from a temporal point of view are transient (i.e., a final unique equilibrium state is achieved), periodic reactions or pulsating flames can arise. Transport processes may also play an important role in these situations, but it is still useful to point out some special characteristics associated with temporal oscillations. Oscillations may arise in an open system having a stable limit cycle. Assuming the transient phase is already past, then the solution of (3) for such a problem would be periodic. This in turn would imply that the inhomogeneity and Jacobian terms in (4) are also periodic functions. Therefore, a condition for resonant matching can be achieved to produce secular type behavior in the sensitivity coefficients. That is, the sensitivity coefficients in an oscillating system may, in fact, not be periodic themselves. More explicitly, this secularity grows out of the simple fact that the frequency $\omega(\alpha)$ of the oscillator will typically be a function of at least some of the system parameters. The concentration profiles in a periodic problem would then be expressed as

$$c_i(t) = \sum_n [a_n^i(\alpha) \cos(n\omega(\alpha)t) + b_n^i(\alpha) \sin(n\omega(\alpha)t)].$$

Differentiation of this expression with respect to a given parameter immediately gives

(8) $$\left(\frac{\partial c_i}{\partial \alpha_j}\right) = \left(\frac{\partial c_i}{\partial \alpha_j}\right)_\omega + \frac{t}{\omega} \frac{\partial \omega}{\partial \alpha_j} \left(\frac{\partial c_i}{\partial t}\right)$$

where the first term on the right-hand side of (8) is a periodic function representing the sensitivity of the cycle with the period held fixed and the last secular term in (8) is directly proportional to the frequency sensitivity coefficient. Without prior removal of the secularity the direct integration of (4) would yield the coefficient on the left-hand side of (8). Since the decomposition on the right-hand side of (8) is quite general, it is possible to separate the computed coefficient into its "structural" and frequency dependent parts. This type of decomposition has been carried out for some kinetic problems, but to the author's knowledge not for the case of combustion systems.

In the case of pure chemistry, the elementary sensitivity coefficients discussed above will likely continue to be the simplest and most useful type of sensitivity information available. An examination of the coefficients gives immediate physical insight into the kinetic processes, and they may also be used to give an estimate of statistical error propagation from the parameters through to the observables (see §2.8).

(2) *Steady state systems.* Steady state combustion systems are technologically important, but in practice they are surely only idealizations since fluctuations of large magnitude may continuously stimulate transient behavior. Nevertheless, the behavior of the steady state limit is important to examine. The relevant equations correspond to the left-hand side of (1) being set to zero and (2a) no longer has any significance. The resultant differential equation therefore represents in general a nonlinear boundary value problem. The modelling equations alone may be solved by two basic routes. First, the full transient equations (1) and (2) may be integrated out to sufficiently long time for stationarity to be achieved. The second procedure would consist of directly dealing with the stationary boundary value problem by an appropriate iterative technique since the equations would typically be nonlinear. These two procedures can be made equivalent since an iterative

algorithm may be expressed in terms of a discretized time transient equation. Consideration of these issues is important for treating the accompanying sensitivity analysis, since numerical implementation of the sensitivity equations can critically depend on the approach to the original modelling system.

As with the purely temporal equations, the stationary sensitivity problem is achieved by taking appropriate partial derivatives of the steady state system

$$\frac{\partial}{\partial x} D_i \frac{\partial}{\partial x} c_i - v \frac{\partial c_i}{\partial x} + R_i(\mathbf{c}, \mathbf{k}) = 0, \quad i = 1, \cdots, N, \tag{9a}$$

$$\mathcal{B}_i \left(\mathbf{c}, \frac{\partial \mathbf{c}}{\partial x} \right) \bigg|_{x_L, x_R} = 0. \tag{9b}$$

Explicit differentiation of this equation will produce the following equations for the elementary sensitivity coefficients:

$$\frac{\partial}{\partial x} D_i \frac{\partial}{\partial x} \left(\frac{\partial c_i}{\partial \alpha_j} \right) - v \frac{\partial}{\partial x} \left(\frac{\partial c_i}{\partial \alpha_j} \right) + \sum_l \left(\frac{\partial R_i}{\partial c_l} \right) \left(\frac{\partial c_l}{\partial \alpha_j} \right)$$

$$= -\frac{\partial}{\partial x} \left(\frac{\partial D_i}{\partial \alpha_j} \right) \frac{\partial}{\partial x} c_i + \frac{\partial v}{\partial \alpha_j} \frac{\partial c_i}{\partial x} - \frac{\partial R_i}{\partial \alpha_j}, \tag{10a}$$

$$\left[\frac{\partial \mathcal{B}_i}{\partial \alpha_j} + \sum_l \left\{ \left(\frac{\partial \mathcal{B}_i}{\partial c_l} \right) \left(\frac{\partial c_l}{\partial \alpha_j} \right) + \left(\frac{\partial \mathcal{B}_i}{\partial (\partial c_l/\partial x)} \right) \left(\frac{\partial^2 c}{\partial x \partial \alpha_j} \right) \right\} \right]_{x_L, x_R} = 0. \tag{10b}$$

Again, the sensitivity equations are linear and inhomogeneous, and this latter structure can be advantageously utilized for achieving their solution [9]. For example, the linearization of the system in (9) will produce exactly the same operator on the left-hand side of (10a). Therefore, a Newton-type algorithm for solving the modelling equations (9a) would operationally have exactly the same form as the sensitivity equations (10a) except that the inhomogeneities would be different. In particular, the inhomogeneities for the Newton algorithm of (9a) would be the residuals at the previous iteration step, and the inhomogeneity in (10a) is explicitly written above. Recognizing that most of the labor in solving these classes of equations resides in generating the system Jacobian $\partial R_i/\partial c_l = J_{il}$ and its inverse, it is then evident that the sensitivity problem should be solvable at little additional expense over that of achieving a solution to the original modelling system. This prospect assumes that meshing of (9a) and (10a) could effectively be the same for acceptable levels of accuracy. Following the discussion in §2.2.1, it should also be possible to carry forth a similar computational algorithm using the transient route to solve both the modelling and sensitivity equations.

The parameters chosen for study in (10) may reside in either the differential equations or the boundary conditions. Parameters residing in the boundary conditions can be particularly interesting on a physical and mathematical basis. For example, in Dirichlet (or Neumann) problems, the actual boundary values (or their slopes) may sometimes be chosen in the laboratory at will. In this case, the boundary condition sensitivities would be measurable in the laboratory. In addition, the degree to which boundary condition variations produce only localized responses in the dependent variables provides a direct measure of the system stability. Boundary condition instability will give rise to large scale sensitivities at substantial distances away from the boundary region.

Some stationary sensitivity results for the model flame system

(11) $$\text{reactant} \xrightarrow{k_1} \text{intermediate} \xrightarrow{k_2} \text{product}$$

are shown in Fig. 4. This problem consists of the linear reaction scheme in (11) along with an accompanying energy conservation equation which has a form similar to that of (9a) for the temperature where the term R_i is replaced by an appropriate energy flux expression. Since the rate constants were taken as having an Arrhenius form, the overall system is still nonlinear. Indeed, in the case of an Arrhenius dependence, the nonlinearity can be so strong as to dominate the entire system sensitivity behavior. The degree to which this is a general phenomenon is not clear at this time, and further extensive numerical calculations would be very helpful in this regard.

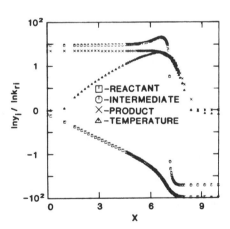

FIG. 4. *Log-normalized sensitivity coefficients for a model flame described by the chemistry (reactant → intermediate → product) and an appropriate energy balance equation [9d]. The system corresponds to a one-dimensional steady premixed flame on the interval $0 \leq x \leq 10$ and y_i is the ith species or the temperature. The "flame" is located near $x \simeq 7$, and the flow is from left to right in the figure. The sensitivities are with respect to the rate constant k_{ri} connecting the reactant and intermediate. It is interesting to observe that sensitivity of the intermediate changes sign in the vicinity of the flame.*

(3) *Transient spatial problems.* The full transient spatial modelling equations are those of (1) and (2). The solution of these equations in even one dimension is a nontrivial task, although such calculations are becoming more routine. In general, the role of modelling in elementary sensitivity analysis discussed in §2.2.1 and 2.2.2 above also apply here. In particular, the method of solution of the modelling equations is of prime concern for achieving an efficient solution of the accompanying sensitivity equations. These latter equations have a structure similar to (10a) except with the addition of a transient time derivative term. Presently there is little experience with numerically dealing with space-time sensitivity systems, but some comments can be made. First, the Green's function solution to the sensitivity equations may be obtained as argued in (5)–(7) above. In the present situation, the Green's function satisfies the following equation

$$(12) \quad \left[1\frac{\partial}{\partial t} - \frac{\partial}{\partial x} \mathbf{D} \frac{\partial}{\partial x} + 1v \frac{\partial}{\partial x} - \mathbf{J} \right] G(x, t; x', t') = 1\delta(x - x')\delta(t - t').$$

In terms of the Green's function matrix the sensitivity coefficients become

$$(13) \quad \frac{\partial c_i(x, t)}{\partial \alpha_j} = \sum_l \int_0^t dt' \int_{x_L}^{x_R} dx' G_{il}(t, t; x', t')$$

$$\times \left[-\frac{\partial}{\partial x'} \frac{\partial D_i}{\partial \alpha_j} \frac{\partial}{\partial x'} c_l(x', t') + \frac{\partial v}{\partial \alpha_j} \frac{\partial c_l(x', t')}{\partial x'} - \frac{\partial R_i}{\partial \alpha_j} \right]$$

when the parameter α_j is not an initial condition parameter. Although the actual calculation of sensitivity coefficients from this route may not be the most practical scheme, it is formally correct and its connection to the subsequent discussion of Green's function behavior will be useful.

A sensitivity analysis of the linear reaction-diffusion scheme

$$(14) \quad c_1 \underset{k_2}{\overset{k_1}{\rightleftarrows}} c_2 \underset{k_4}{\overset{k_3}{\rightleftarrows}} c_3$$

has been thoroughly analyzed both numerically and analytically [10]. The linearity of the system allows for a complete analytical treatment as a test of numerical coding. In practice, the calculations on this problem were mainly confined to functional sensitivity analysis whereby the parameters (rate constants and diffusion coefficients) were treated as being time and space dependent. Therefore, a further discussion of these results is reserved for §2.3 below. In general, sensitivity analysis of space-time modelling represents a current forefront topic both theoretically and in practical numerical terms. It is anticipated that advances in this area will occur in the coming years.

2.3. Functional sensitivity analysis. Thus far, all of the parameters entering into the differential equations were treated as constants both with regard to their nominal values as well as for any variations. In some circumstances, the nominal values of the system parameters may actually be space and/or time dependent, and clearly an extension of the elementary sensitivity concepts above is necessary. In addition, even if the nominal parameter values are themselves constant, it is not necessary to confine variations to being constant. In this regard, functional sensitivity analysis represents a significant generalization of constant parametric sensitivity analysis. In order to proceed with this extended formulation, it is first necessary to define appropriate functional derivatives. Both Gâteux and Fréchet type derivatives have been considered. We shall confine our attention here to the Fréchet type since they are rather convenient for computational purposes. Therefore, the functional variation of the ith concentration is expressed as

$$(15) \quad \delta c_i(x, t) = \sum_l \int dt' \int dx' \frac{\delta c_i}{\delta \alpha_j}(x, t; x', t') \delta \alpha_j(x', t')$$

where $\delta \alpha_j(x', t')$ is an arbitrary incremental functional variation of the jth system parameter. Analogous definitions to (15) would apply to the functional variations of other quantities as they arise in the analysis. Therefore, proceeding in a rather straightforward manner we may functionally differentiate (1) and (2) with respect to any arbitrary system parameter. The result of this variation will produce the following differential equations for the elementary functional sensitivity densities

$$(16) \quad \left[\frac{\partial}{\partial t} - \frac{\partial}{\partial x} D_i \frac{\partial}{\partial x} + v \frac{\partial}{\partial x}\right] \frac{\delta c_i}{\delta \alpha_j}(x, t; x', t') - \sum_l \frac{\partial R_i}{\partial c_l} \frac{\delta c_l}{\delta \alpha_j}(x, t; x', t')$$
$$= -\frac{\partial}{\partial x}\left(\frac{\delta D_i}{\delta \alpha_j}\right)\frac{\partial}{\partial x} c_i + \frac{\delta v}{\delta \alpha_j}\frac{\partial}{\partial x} c_i + \frac{\delta R_i}{\delta \alpha_j},$$

$$(17a) \quad \frac{\delta c_i}{\delta \alpha_j}(x, 0; x', t') = \begin{cases} 0, & \alpha_j \neq c_i(x, 0), \\ \delta(x - x')\delta(t'), & \alpha_j = c_i(x, 0), \end{cases}$$

$$(17b) \quad \left[\frac{\delta \mathcal{B}_i}{\delta \alpha_j}(x, t; x', t')\right.$$
$$+ \sum_l \left\{\left(\frac{\partial \mathcal{B}_i}{\partial c_l}\right)\frac{\delta c_l}{\delta \alpha_j}(x, t; x', t')\right.$$
$$+ \left.\left.\left(\frac{\partial \mathcal{B}_i}{\partial(\partial c_l/\partial x)}\right)\frac{\partial}{\partial x}\frac{\delta c_l}{\delta \alpha_j}(x, t; x', t')\right\}\right]\Bigg|_{x=x_L, x_R} = 0.$$

The actual detailed structure of the right-hand side of (16) and the derivative $\delta \mathcal{B}_i(x, t; x', t')/\delta \alpha_j$ depend on the particular choice of parameter α_j. A thorough exposition of this matter can be found in the literature [10]. It is apparent that these equations have the same linear inhomogeneous character as found in §2.2 above. Therefore, much of the same general discussion follows through about their solution. The resultant coefficients $\delta c_i(x, t; x', t')/\delta\alpha_j$ which enter into (15) have the interpretation of the response of the ith species concentration at position x and time t with regard to a disturbance of the jth parameter at position x' and time t'. Due to causality, these responses will always be zero for $t' > t$. These coefficients clearly contain very detailed information on system response. In situations where parameter disturbances may actually be made at selected points in space and/or time, knowledge of these gradients can be especially valuable for estimating system control characteristics. In the other extreme, where the system parameters are constant along with the parameter variations, it is still possible to calculate the sensitivity density coefficients (functional derivatives). Furthermore, in the latter circumstances it is apparent from the structure of (16) and (17) that an integration over x' and t' will reduce the equations to those of ordinary parametric sensitivity analysis. Therefore, we can establish the validity of the following integral relationship for the case where the nominal system parameters are constants

$$(18) \qquad \frac{\partial c_i}{\partial \alpha_j}(x, t) = \int_0^t dt' \int_{x_L}^{x_R} dx' \frac{\delta c_i}{\delta \alpha_j}(x, t; x', t').$$

With the relation in (18), one may argue that calculation of the functional sensitivity densities will always be preferable on physical grounds since they inherently contain more information than the sensitivity coefficients, and the latter quantities may always be obtained by integrating the densities over space and time when appropriate. However, practical computational issues can intercede into this goal.

Direct numerical integration of (16) and (17) has not been accomplished at this time for any combustion system. The Green's function analysis discussed in §2.1 above may also be incorporated here. For example, assuming that α_j is a particular rate constant entering into the reaction term of (1), it is evident that the right-hand side of (16) simply becomes $\delta R_i/\delta\alpha_j = \partial R_i/\partial k_j \cdot \delta(x - x')\delta(t - t')$. Therefore, utilizing this observation along with the Green's function defined in (12), it follows that the sensitivity density from (16) has the simple structure

$$(19) \qquad \frac{\delta c_i}{\delta k_j}(x, t; x', t') = \sum_l G_{il}(x, t; x', t') \frac{\partial R_l}{\partial k_j}(x', t')$$

where the coordinate and time dependence of $\partial R_l(x', t')/\partial k_j$ enters implicitly through **c** and **k**. The relationship in (19) again shows the fundamental role played by the system Green's function. In the present circumstances, the functional derivative with respect to a rate constant is simply the Green's function weighted by the parametric derivative of the reaction terms in the kinetic equations. Typically, the latter derivative will only be nonzero for relatively few terms to further simplify the summation in (19).

Although numerical integration of (16) has not been achieved, the formalism has been applied to an analytically soluble reaction-diffusion system with chemical terms described by (14) [10]. Even this simple system has exhibited a rather rich variety of behavior, and one can only imagine the interesting structure that must lie in more realistic systems. For the sake of illustration, Fig. 5 shows the functional derivative $\delta c_1(t, t')/\delta k_4$ corresponding to the reduced case of pure temporal kinetics. The left quadrant of the figure is zero for reasons of causality and an effective time delay is present indicating that a disturbance of k_4 at time t' requires a finite propagation time for c_1 to be maximally influenced. For more complex systems involving lengthy convoluted pathways between chemical species, parametric response surfaces having longer delays and more detailed structure would likely arise. Much could be learned about chemical kinetic systems, including those of combustion, by examination of functional sensitivity analysis, and a careful development of numerical implementation procedures is needed.

2.4. Stochastic systems. The discussion thus far has assumed that all of the independent and dependent variables in the system models were deterministic in nature. The parameters entering into (1) and (2) may have a probability distribution function reflecting the uncertainty in their values. Under these conditions the natural goal is to seek the resultant probability distribution function of the concentrations. Equations (1) and (2), when cast

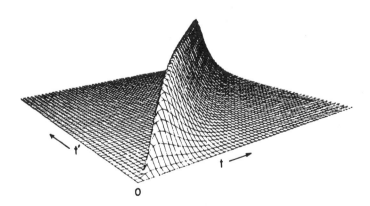

FIG. 5. *Functional derivative $\delta c_i(t, t')/\delta k_4$ from the temporal system with chemistry described by (14). Due to causality, the sensitivity density is zero for $t < t'$ and the maximum sensitivity response occurs in the vicinity $t - t' \simeq 0.2$. This latter time difference corresponds to the finite time delay needed for the disturbance of k_4 to propagate and maximally influence the first chemical species* [10].

in this form, should be thought of as stochastic differential equations. Stochasticity may also arise due to inherent fluctuations of the dependent variables. For example the temperature may fluctuate and thereby cause temperature dependent parameters in turn to fluctuate. In addition, it must be realized that (1) and (2) are in fact continuum approximations to a more proper discrete physical picture. The fluctional nature of the concentrations may also be treated by considering (1) as stochastic and invoking the fluctuation dissipation theorem [11]. Again, one would seek the probability distribution function of the concentrations in addition to appropriate sensitivity coefficients. Stochastic kinetic phenomena have received increasing interest, and an extensive literature is available [12]. The sensitivity analysis of these problems has only been studied rather sparsely and just for the case of purely temporal systems. Therefore the discussion below will be confined to this latter category of problems.

Regardless of the physical origin of the stochasticity, the appropriate quantity of interest from the modelling will be the probability distribution function for the concentrations (or other dependent variables, as appropriate). The distribution function $P(\mathbf{c}, t)$ may be used to calculate the expectation value of any meaningful physical observables

$$\langle g \rangle = \int P(\mathbf{c}, t) g(\mathbf{c}) \, d\mathbf{c} \tag{20}$$

where $g(\mathbf{c})$ is an appropriate function whose expectation value is sought. In practice, the mean $\langle \mathbf{c} \rangle$ and covariance $\text{cov}(\mathbf{c}, \mathbf{c})$ are often used as simple characteristics of the probability distribution function

$$\langle c_i \rangle = \int c_i P(\mathbf{c}, t) \, d\mathbf{c}, \tag{21a}$$

$$\text{cov}(c_i, c_j) = \int (c_i - \langle c_i \rangle)(c_j - \langle c_j \rangle) P(\mathbf{c}, t) \, d\mathbf{c}. \tag{21b}$$

Issues related to this matter will also be treated in §2.8 below. Sensitivity information would be achieved by differentiating (20) or (21) with respect to any relevant system parameters. When these latter parameters are themselves fluctuating quantities, it would then be meaningful to consider the proper parameters as the mean and/or covariance matrix elements of the probability distribution function of the parameters.

For the purely temporal case of (1), its stochastic extension would have the following form as a generalized Langevin equation

$$\frac{dc_i}{dt} = R_i(\mathbf{c}, \mathbf{k}) + \sum_j Q_{ij}(\mathbf{c}, \mathbf{k}) \xi_j(t) \tag{22}$$

where $\xi(t)$ is a delta correlated Gaussian stochastic process,

$$\langle \xi_i(t) \rangle = 0, \qquad \langle \xi_i(t) \xi_j(t') \rangle = \delta_{ij} \delta(t - t'). \tag{23}$$

This process corresponds to white noise, and other situations characterized by colored noise correlation functions could also be treated. The matrix **Q** entering into (22) may depend explicitly on the species concentrations which would correspond to a stochastic differential equation having multiplicative noise. In addition **Q** might also depend on other parameters besides its implied rate constant dependence in (22). In the case of concentration fluctuations, one may argue the form of **Q** by the fluctuation dissipation theorem which effectively relates its elements to the forward and reverse reaction rates entering into the vector **R**. Equation (22) applies to the cases where the concentrations or parameters are inherently fluctuating quantities and the stochastic variable $\xi(t)$ would have an appropriate corresponding meaning. In the situation where there are no fluctuations then **Q** would not be present but the rate constants might still have statistically distributed values. It is important to recognize that in this latter case the kinetic equations are deterministic, although the probability distribution function of the concentrations is still sought. Fluctuation phenomena giving rise to the stochastic term in (22) could be especially important in certain combustion situations. In particular, systems capable of chemical instabilities (e.g., an explosion) could be extremely sensitive to fluctuation phenomena near the critical onset of the unstable behavior. Accordingly, sensitivity coefficients in these problems should show dramatic alterations near the critical points and in the presence of fluctuations.

In order to properly interpret (22), it is necessary to specify the statistical sense in which the equation is to be integrated. On physical grounds, this stochastic differential equation is interpreted in the sense of Stratonovich, and the reader is referred to the literature for further information concerning this point [11], [12]. This system of differential equations would then be integrated by sampling an appropriate number of trajectories. Sensitivity equations corresponding to (22) may be derived in the usual fashion by merely differentiating the equation with respect to a chosen system parameter (i.e., either a rate constant or initial condition). The resultant differential equation will as usual be linear and inhomogeneous. It is possible to generate sensitivity coefficients of the expectation value in (20) having the following form

(24a) $$\frac{\partial}{\partial \alpha_j} \langle g \rangle = \lim_{\epsilon \to 0} \sum_i \int d\tau \langle g(\mathbf{c}) \xi_i(\tau + \epsilon) b_{ij}(\mathbf{c}(\tau), \boldsymbol{\xi}(\tau), \boldsymbol{\alpha}) \rangle$$

where

(24b) $$b_{ij} = \sum_n Q_{in}^{-1} \left[\delta_{\alpha_j, c_{j(0)}} \delta(\tau) + \frac{\partial R_n}{\partial \alpha_j} + \sum_l \frac{\partial Q_{nl}}{\partial \alpha_j} \xi_l(\tau) \right].$$

It is interesting to note that the closed form expression for the sensitivity in (24) actually does not involve the system Green's function in contrast to the deterministic case in (7). In practice, the evaluation of the integral in (24) would be achieved by appropriate statistical sampling of concentration trajectories. However, a practical limiting form for (24) can be obtained by performing a linearization about the relevant nonstochastic concentration trajectory. This linearization is based on the assumption that the magnitude of the concentration fluctuations are relatively small compared to the mean concentrations themselves. The interesting point here is that within this framework, the mean concentrations, the concentration correlation functions and their respective sensitivities may be obtained exclusively in terms of deterministic trajectory information. Therefore, within the framework of the same computational procedures presented in §2.2, one may also obtain information on the significance of fluctuations through appropriate sensitivity coefficients.

A Fokker–Planck equation may be derived for the probability distribution function $P(\mathbf{c}, t)$ by first noting that

$$P(\mathbf{c}, t) = \langle \delta(c_1 - \tilde{c}_1(t)) \cdots \delta(c_N - \tilde{c}_N(t)) \rangle \tag{25}$$

where it is understood that the vector $\tilde{\mathbf{c}}(t)$ is the actual solution to (22) and therefore depends on the stochastic variables. Differentiation of (25) with respect to time followed by utilization of (22) will finally produce the desired Fokker–Planck equation for the probability distribution function

$$\frac{\partial}{\partial t} P(\mathbf{c}, t) = -\nabla_c \cdot (\mathbf{R}P) + \tfrac{1}{2}\nabla_c \cdot (\mathbf{Q} \cdot \nabla_c(\mathbf{Q}P)) \tag{26}$$

where ∇_c is the gradient operator acting in the space of concentration variables. Under the circumstance that $\mathbf{Q} = 0$, the last term in (26) is not present [3], and the resultant equation would be appropriate for situations where the concentrations are not fluctuating and the otherwise constant parameters have a simple statistical distribution.

Sensitivity analysis in systems actually requiring stochastic simulations has not been carried out at this time. However, calculations have been performed within the framework of the linearized approximation [13]. In particular, the system

$$c_1 + 2c_2 \rightleftarrows 3c_2, \tag{27}$$

$$c_2 \rightleftarrows c_3$$

has been studied as a model for explosive behavior induced by the autocatalytic first step of the reaction. Explosive behavior is indicated by a trajectory having the characteristics $c_2(t) > c_2(0)$ for some time $t > 0$. When a nonstochastic trajectory violates the criteria for an explosion, the presence of fluctuations can nevertheless induce an explosion. In a similar vein, when a deterministic trajectory satisfies the explosive criteria, the presence of fluctuations can induce *stability* in the sense that there may be a finite probability for nonexplosive behavior in that regime when fluctuations are present. The sensitivity of the probability distribution function and various other responses may similarly be calculated to indicate which parameters are controlling the enhancement or dehancement of explosive behavior.

2.5. Green's functions and stability analysis. The Green's function matrix was introduced in (6) and (12) as a means for solving the elementary sensitivity equations. At this point, we will emphasize the physical and mathematical meaning of the Green's function. It will be shown that the Green's function matrix elements are just a specialized class of elementary sensitivity coefficients themselves. In addition, knowledge of the Green's function provides valuable information regarding the stability of the kinetic system. Stability information can be extremely important in combustion problems, particularly those that are capable of explosive behavior.

The simplest understanding of the Green's function may be achieved by first considering the temporal system in (3). Differentiation of these equations with respect to the concentration $c_j(t')$, $t' \leq t$ will produce the following differential equations:

(28a) $$\frac{\partial}{\partial t}\left(\frac{\partial c_i(t)}{\partial c_j(t')}\right) - \sum_l \left(\frac{\partial R_i}{\partial c_l}\right)\left(\frac{\partial c_l(t)}{\partial c_j(t')}\right) = 0,$$

(28b) $$\left.\frac{\partial c_i(t)}{\partial c_j(t')}\right|_{t=t'} = \delta_{ij}.$$

It is evident that (28a) has exactly the same form as (6) for $t > t'$. If in addition, causality is imposed such that $\mathbf{G}(t, t') = \mathbf{0}$, $t < t'$, then an integration of (6) over t around the neighborhood of the initial time t' will produce the initial condition $\mathbf{G}(t', t') = \mathbf{1}$. This latter condition is identical with (28b), and therefore we may conclude that the Green's function in (6) has the following interpretation

(29) $$G_{ij}(t, t') = \left(\frac{\partial c_i(t)}{\partial c_j(t')}\right).$$

In the temporal case these Green's function elements may be interpreted as the sensitivity of the ith chemical species at time t with respect to a disturbance of the jth chemical species at time t'. In other words, these elements are initial condition sensitivities and since the equations are temporal, a solution at any time t' will be the initial condition for the solution after that time. In addition, the connection of these gradients to stability analysis and control problems is also evident, and this issue will be discussed further below.

Beyond just the purely temporal case, a more general derivation of the Green's function elements may be arrived at by considering (1) again. This latter equation will now be modified by the addition of an (incremental) flux term $\mathcal{J}_i$ as a source for the ith equation

$$(30) \qquad \frac{\partial c_i}{\partial t} = -v \frac{\partial c_i}{\partial x} + \frac{\partial}{\partial x} D_i \frac{\partial}{\partial x} c_i + R_i(\mathbf{c}, \mathbf{k}) + \mathcal{J}_i(x, t).$$

Functional differentiation of (30) with respect to the new added flux terms $\delta/\delta\mathcal{J}_i(x', t')$ will produce

$$(31) \qquad \left[\mathbf{1} \frac{\partial}{\partial t} - \frac{\partial}{\partial x} \mathbf{D} \frac{\partial}{\partial x} + \mathbf{1}v \frac{\partial}{\partial x} - \mathbf{J} \right] \frac{\delta \mathbf{c}}{\delta \mathcal{J}}(x, t; x', t') = \mathbf{1}\delta(x - x')\delta(t - t')$$

where the matrix of dependent variables has elements

$$(32) \qquad \left(\frac{\delta \mathbf{c}}{\delta \mathcal{J}}(x, t; x', t') \right)_{ij} \equiv \frac{\delta c_i}{\delta \mathcal{J}_j}(x, t; x', t').$$

Functional differentiation of (2a) and (2b) will produce appropriate initial and boundary conditions for $\delta c_i/\delta\mathcal{J}_j$. Comparison of (31) and (12) will immediately lead to the identification

$$(33) \qquad G_{ij}(x, t; x', t') = \frac{\delta c_i}{\delta \mathcal{J}_j}(x, t; x', t')$$

such that the i-j element of $\mathbf{G}$ has the interpretation as the functional variation of the ith species at position x and time t with respect to a disturbance of the jth species flux at position x' and time t'. The special case of purely temporal kinetics may be arrived at by removing the spatial derivative terms from (31) and integrating the equation over all space. In a similar fashion, a steady state Green's function may also be identified as having elements $\delta c_i(x, x')/\delta\mathcal{J}_j$ with a similar interpretation. In the case of pure temporal systems, it may be shown that the Green's function satisfies a group multiplication property

$$(34) \qquad \mathbf{G}(t'', t) = \mathbf{G}(t'', t')\mathbf{G}(t', t), \qquad t'' \geq t' \geq t$$

while in the space-time case, the nonlocal nature of the spatial coordinate produces an analogous integral relationship

$$(35) \qquad \mathbf{G}(x'', t''; x, t) = \int dx' \mathbf{G}(x'', t''; x', t')\mathbf{G}(x', t'; x, t).$$

Equations (34) and (35) allow the Green's functions to be interpreted as fundamental propagators for the system sensitivity behavior.

The Green's function matrix elements in (29) and (33) are unique among sensitivity coefficients (except for sensitivities with respect to actual control variables), in that they are in principle measurable in the laboratory. This measurement would be achieved by disturbing a given species and monitoring the response amongst all of the chemical species of the system. In this regard, there is clearly a distinct difference between temporal disturbances and their spatial analogs. In particular, due to causality a disturbance at time t' can only affect future times while a disturbance at point x' could affect the system throughout space. The latter comment is important in flame systems since a disturbance downstream in a flow could affect the upstream behavior particularly for unstable problems. These interpretations of the Green's function elements immediately lead to the issue of system control. In particular, it may be possible to utilize the Green's functions as a guide to the alteration of spatial and temporal behavior by the judicious introduction of chemical species or other dependent control variables.

In a sense, the Green's function matrices have a fundamental role in sensitivity analysis since all system sensitivities may be expressed as linear combinations of the Green's function elements. This comment is evident from the structure of (7) and (13). Therefore, although these latter equations may not always be utilized in practice for the computation of elementary sensitivities, an examination of the Green's function elements can be extremely informative for providing a broad understanding of sensitivity behavior. Another significant point to emphasize is that the dimension of the Green's function matrix is $N \times N$, regardless of the number of system parameters M. An interesting related issue concerns the degree to which the individual matrix elements of **G** appear as independent functions of their time and/or spatial arguments. A number of calculations have indicated that portions of the Green's function matrix often have quite similar behavior. An example of this type is shown in Fig. 6 from the oxidation of carbon monoxide in the presence of water [8]. In the present case, the similar Green's function curves result from the strong coupling between the radical species in the reaction. It appears that strong coupling of various types can give rise to similar Green's function matrix element behavior. This point is important since it bears on the issue of how many truly independent species exist in the reacting mixture. Another case is illustrated in Fig. 7 corresponding to the reaction-diffusion system having the chemistry given by (14). The figure clearly illustrates the points made above concerning how a local disturbance propagates through the chemical system. In a closed chemical reactor corresponding to the cases of Figs. 6 and 7, it is evident that the system will return to its former behavior at a sufficiently long time after the initial disturbance. For open systems this may not always be the case and permanent alteration of the final product profiles could occur due to flux disturbances.

The comments at the end of the last paragraph naturally lead to the consideration of physical stability in combustion and kinetic problems. The sig-

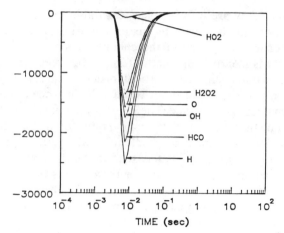

FIG. 6. *Green's function matrix elements $\partial CO(t)/\partial X(O)$ where the species X are indicated on the figure. The system corresponds to the oxidation reactions involving the chemistry of $CO/O_2/H_2O$. At first, there is no apparent response to the radical disturbances followed by an active period and ultimately decay back to former system behavior. The similarity between the shapes of the curves is the result of the strong coupling between the chemical radicals.*

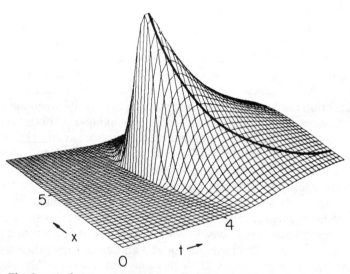

FIG. 7. *The Green's function matrix element $\partial c_1(x, t; 5, 4)/\partial \mathcal{J}_2$ for the one-dimensional reaction-diffusion system having chemistry described by (14) [10]. The disturbance of the second chemical species occurs at the location $x' = 5$ and $t' = 4$. The solid line is the locus of maxima for cuts through the surface at constant x, and it indicates a longer response delay of the first chemical species for measurements further away from the disturbance point $x' = 5$.*

nificance of the Green's function matrix for this matter is evident from the interpretation given to (29) and (32). Chemical or physical instability would be manifested by large magnitude Green's function matrix elements. A runaway chemical reaction would have the characteristic that a small initial disturbance would eventually cause enormous growth to occur. A precise way of quantifying this issue can be achieved by diagonalizing the Green's function matrices. As an illustration, consider just the purely temporal case where this diagonalization will take on the form

$$(36) \qquad \mathbf{g}(t, t') = \mathbf{U}^{-1}(t, t')\mathbf{G}(t, t')\mathbf{U}(t, t')$$

with $\mathbf{g}$ being a diagonal matrix of eigenvalues and the columns and rows of $\mathbf{U}$ and $\mathbf{U}^{-1}$, respectively, forming the right- and left-hand eigenvectors of $\mathbf{G}$. In stability analysis, it is traditional to define the asymptotic Lyapunov coefficients as

$$(37) \qquad \lambda_i = \lim_{t \to \infty} \frac{1}{t} \ln g_i$$

where a positive real part to λ_i will indicate unstable behavior due to system growth. The asymptotic analysis based on (37) is not appropriate if the short time scale behavior for a system is of interest. For example, a closed reactor containing fuel and oxidizer will always exhibit asymptotic thermodynamic stability while its transient stability behavior is the topic of actual concern. Therefore, the time dependence of the eigenvalues $g_i(t, t')$ is of more concern in many problems. In addition, the eigenvectors in (36) may be interpreted as prescribing particular linear combinations of initial or final chemical species associated with the characteristic growth constants (eigenvalues). For example, in a combustion system, if $g_i(t, t')$ indicated explosive growth (i.e., $|g_i(t, t')| > 1$), then the eigenvector components would identify which species were participating to produce this behavior. An analysis of this type has been carried out for the reaction scheme in (27) which is capable of explosive growth under appropriate conditions [14].

For the reasons cited above, it is generally recommended that the Green's function matrix be calculated when performing any sensitivity analysis. The Green's function matrix may be available routinely as a result of some numerical schemes, and in others the added cost will typically be minimal. As well as providing insight into general parametric system sensitivity, the Green's function elements have their own physical significance.

2.6. Feature and objective function sensitivity analysis. Thus far in the discussion, the observables of concern for sensitivity analysis were the natural dependent variables (i.e., species concentrations or the temperature). In realistic problems, particularly of a practical nature, it may be more natural to consider the desired output as a function or functional of the chemical species. This could happen, for example, when thermodynamic properties are of importance in combustion or other kinetic systems of concern. Another common circumstance can arise when examining the concentration or temperature profiles as a function of the independent coordinates or time. For example, the time history of a concentration may typically start out having a small value, rise to an intermediate extremum and again decrease as the reaction proceeds. Employment of the techniques in §2.2 would allow for calculating sensitivities $\partial c_i(t)/\partial \alpha_j$ at any time along the curve. However, given the simple characteristic behavior cited above, more interesting questions may concern which parameters control the width or location of the concentration extremum valve. In other problems, the parameters controlling the location of a threshold or critical slope may also be appropriate. It is convenient to consider the output to have its own characteristic parameters β_1, β_2, $\cdots$ which will be distinct from the input parameters α_1, α_2, $\cdots$ entering into the physical model. The output parameters could be thought of as desired objectives or simple characteristics associated with concentration profiles.

The distinction between input (α) and output (β) parameters should be thought of as a natural one in kinetic systems. This comment follows from two considerations. First, although the input parameters would typically have a very physical role in the mechanism or transport processes, their actual role in the physical output can be totally obscured by the intervening differential equations. This issue is exactly the point made in Fig. 1. On the other hand, the experimental data or objectives of interest have their own natural parameters which may be identified without having set up the mathematical model for the physical process. This point of view is particularly relevant when considering the inversion of experimental data since from this perspective, the output is just as significant as the input and neither should be thought of as more fundamental than the other. From a modelling point of view, one would of course start with the input parameters and proceed forward through to the output. In this case, it is natural to seek an understanding of the function relationship

$$(38) \qquad \beta = \beta(\alpha).$$

A similar functional expression will arise for cases when the parameters are spatially or temporally distributed. A full understanding of the relation in (38) would constitute the very difficult task of global mapping. More realistically, we may seek corresponding sensitivity information $\partial \beta_i/\partial \alpha_j$ as a probe of the relationship in (38). Two approaches to seeking sensitivity information of this type consist of fitting techniques and feature (objective) function sensitivity analysis. These approaches will be discussed below.

The fitting technique is in principle straightforward and may be illustrated for a temporal problem as follows. Consider the ith species concentration

$c_i(t, \boldsymbol{\alpha})$ as obtained from the mathematical modelling. From an examination of the time dependence of this function a choice of an explicit functional form $\tilde{c}_i(t, \boldsymbol{\beta})$ would be made for fitting purposes. A minimizing functional of least squares (or other form) would then be introduced

$$(39) \qquad R = \int_{t_1}^{t_2} dt [c_i(t, \boldsymbol{\alpha}) - \tilde{c}_i(t, \boldsymbol{\beta})]^2$$

where the integral is over the interval $t_1 \leq t \leq t_2$ containing the feature (objectives) of interest. Weighted functionals can also be used in order to emphasize a given sub-interval of time. Minimization of the residual in (39) with respect to the parameters $\beta_1, \beta_2, \cdots$ will imply the relationship in (38). It is of course recognized that the actual model is run with the input parameters $\boldsymbol{\alpha}^0$ set at their nominal values. Therefore, differentiation of (39) first with respect to $\beta_1, \beta_2, \cdots$ followed by differentiation with respect to $\alpha_1, \alpha_2, \cdots$ will yield the desired sensitivity coefficients

$$(40a) \qquad \frac{\partial \beta_n}{\partial \alpha_l} = (\mathbf{C}^{-1} \mathbf{D})_{nl}$$

where

$$(40b) \qquad C_{jn} = \int_{t_1}^{t_2} dt \left\{ \left(\frac{\partial \tilde{c}_i}{\partial \beta_n}\right)\left(\frac{\partial \tilde{c}_i}{\partial \beta_j}\right) - [c_i - \tilde{c}_i] \left(\frac{\partial^2 \tilde{c}_i}{\partial \beta_j \partial \beta_n}\right) \right\},$$

$$(40c) \qquad D_{jl} = \int_{t_1}^{t_2} dt \left(\frac{\partial \tilde{c}_i}{\partial \beta_j}\right)\left(\frac{\partial c_i}{\partial \alpha_l}\right).$$

The only input information needed from the mathematical model is the concentration profile and its sensitivity coefficients; the remaining quantities in (40) involve the fitted function and its derivatives. This somewhat pedestrian technique is only limited by one's ability to choose reasonable functional forms. In addition it has another especially valuable potential contribution in the area of parameter scaling. In particular, knowledge of the gradients in (40) would allow for an estimate of the variation of the feature parameters with respect to the input parameters

$$(41) \qquad \beta_i \simeq \beta_i(\boldsymbol{\alpha}^0) + \sum_l \left(\frac{\partial \beta_i}{\partial \alpha_l}\right)(\alpha_l - \alpha_l^0).$$

Although the expansion in (41) is only taken to linear order, the dependence of $\tilde{c}_i(t, \boldsymbol{\beta})$ on β_i would typically be very nonlinear. Therefore, it may be possible to develop nonlinear scaling formulas with respect to variations in $\boldsymbol{\alpha}$ by utilizing the expansion in (41), $c(t, \boldsymbol{\alpha} + \Delta\boldsymbol{\alpha}) \simeq \tilde{c}[t, \boldsymbol{\beta}(\boldsymbol{\alpha} + \Delta\boldsymbol{\alpha})]$.

Another form of feature sensitivity analysis may be realized by recognizing that any meaningful physical or mathematical feature will generally be expressible as a functional of the concentration

$$\beta = \int_{t_1}^{t_2} F[\mathbf{c}(t, \boldsymbol{\alpha}), t]\, dt \tag{42}$$

where F is an appropriate function. In some cases, this latter function may involve Dirac delta functions which would eliminate the integral form of (42). In other cases, the relationship implied by this equation and the feature parameter may be implicit. For example, suppose that the sensitivity of the location of a concentration extremum is sought. The concentration extremum is defined by the equation

$$\frac{\partial c}{\partial t}[t(\boldsymbol{\alpha}), \boldsymbol{\alpha}]\bigg|_{t=t^*} = 0 \tag{43}$$

where it is understood that the derivative is evaluated at the extremum locaction t^*. Equation (43) implies a relationship between t^* and the system parameters. Therefore, differentiation of (43) with respect to the ith parameter will produce the following results:

$$\left(\frac{\partial t^*}{\partial \alpha_i}\right) = -\left(\frac{\partial^2 c}{\partial t \partial \alpha_i}\right) \bigg/ \left(\frac{\partial^2 c}{\partial t^2}\right). \tag{44}$$

Feature sensitivity coefficients such as these may be derived for any identifiable pointwise characteristics of a concentration profile [15]. A thorough analysis of this type was carried out for the moist oxidation of carbon monoxide for the purpose of elucidating which parameters control the shape and structure of the concentration profiles [8].

Returning again to (42), we may differentiate it with respect to the jth system parameter and obtain

$$\frac{\partial \beta}{\partial \alpha_j} = \sum_l \int_{t_1}^{t_2} \left(\frac{\partial F(t)}{\partial c_l}\right)\left(\frac{\partial c_l(t)}{\partial \alpha_j}\right) dt \tag{45}$$

where for simplicity of notation only the t-dependence is shown in the functions of the integrand. As an example, we may consider the parameter α_j as not being an initial condition, and application of (7) will yield the sensitivity coefficients needed in the integrand of (45).

$$\frac{\partial c_l(t)}{\partial \alpha_j} = \sum_{l'} \int_0^\infty G_{ll'}(t, t') \frac{\partial R_{l'}(t')}{\partial \alpha_j} dt' \tag{46}$$

where the upper limit of the integral has been taken to infinity without any approximation since the Green's function is zero for $t' > t$. Equations (46) and (45) may now be combined into the form

$$\frac{\partial \beta}{\partial \alpha_j} = \sum_{l'} \int_0^\infty dt'\, g_{l'}(t') \frac{\partial R_{l'}(t')}{\partial \alpha_j} \tag{47a}$$

where

(47b) $$g_{l'}(t') = \sum_l \int_{t_1}^{t_2} dt \, \frac{\partial F(t)}{\partial c_l} G_{ll'}(t, t').$$

Equation (47a) is computationally quite practical when it is recognized that the reduced Green's function $g_{l'}(t')$ in (47b) may be calculated without first calculating the full Green's function matrix $\mathbf{G}$ in its integrand. This observation is based on the fact that we may define the adjoint Green's function $\mathbf{G}^\dagger(t', t) = \mathbf{G}(t, t')$ to satisfy the differential equation

(48) $$\frac{\partial}{\partial t'} \mathbf{G}^\dagger(t', t) + \mathbf{G}^\dagger(t', t)\mathbf{J}(t') = \mathbf{1}\delta(t - t')$$

with $t' \leq t$ and $\mathbf{G}^\dagger(t, t) = \mathbf{1}$. Therefore, the reduced Green's function satisfies the equation

(49) $$\frac{\partial}{\partial t'} \mathbf{g}(t') + \mathbf{g}(t')\mathbf{J}(t') = \begin{cases} \partial F(t')/\partial \mathbf{c}, & \text{if } t' \leq t_2, \\ 0, & \text{otherwise.} \end{cases}$$

The savings implied by this approach is apparent in (49) where we need to only solve a vector differential equation rather than the matrix sensitivity differential equations for $\mathbf{G}$. This approach becomes attractive when objective functions of the form of (42) may be defined *prior* to actually performing the sensitivity analysis. When the objectives are not understood prior to evaluating the system, then it will be necessary to calculate the sensitivity coefficients entering into (45) by standard means. In certain engineering applications of combustion processes the objectives of interest may be known beforehand, and the adjoint technique would then be attractive. At this point, applications of this type have not been carried out in combustion, although they have been performed in other problems [15a].

In general terms, feature or objective sensitivity analysis may be one of the most practical applications of the subject. This conclusion is based on the fact that features or objectives are often the only meaningful quantities in the laboratory. In addition, focusing on a reduced set of output characteristics can possibly render system sensitivity information into its most essential form.

2.7. Derived sensitivity analysis. The formulation of modelling and sensitivity analysis encompassed by Fig. 1 implies a definite set of dependent and independent variables. In particular, the species concentrations (or the temperature) were the dependent variables and the parameters in the model were treated as independent variables. Therefore, within the framework of first order sensitivity analysis, a differential change in the concentrations would be expressed in terms of a corresponding differential change in the parameters

$$(50) \quad dc_i(t) = \sum_j \left(\frac{\partial c_i(t)}{\partial \alpha_j}\right) d\alpha_j$$

where the first order elementary sensitivity coefficients provide the connection between these two classes of variables. The time t plays the role of an implicit variable and a similar expression to (50) could be written down for the more general space-time situation. In addition, (15) is exactly the analogous expression corresponding to (50) in the case of functional variations. For the sake of simplicity here, we shall focus on the constant parametric case of (50).

Derived sensitivities are arrived at by considering a rearrangement of the dependent and independent variables as depicted in (50). This could happen a number of different ways, and without loss of generality we may suppose that the first s concentrations are to be transformed into independent variables and the corresponding set of first s parameters are to become dependent variables. A transformation of this type is of the same form as performed routinely in thermodynamics for similar purposes. The physical significance of these transformations will be discussed after presenting the mathematical manipulations giving rise to them. It is convenient at this point to rewrite (50) in matrix-vector notation

$$(51) \quad \begin{pmatrix} dc_1 \\ \vdots \\ dc_s \\ dc_{s+1} \\ \vdots \\ dc_N \end{pmatrix} = \begin{pmatrix} \mathbf{S}_1 & \vdots & \mathbf{S}_2 \\ \cdots & \vdots & \cdots \\ \mathbf{S}_3 & \vdots & \mathbf{S}_4 \end{pmatrix} \begin{pmatrix} d\alpha_1 \\ \vdots \\ d\alpha_s \\ d\alpha_{s+1} \\ \vdots \\ d\alpha_M \end{pmatrix}$$

where the sensitivity matrix $\mathbf{S}$ having elements $S_{ij} = \partial c_i / \partial \alpha_j$ has been partitioned into four blocks $\mathbf{S}_1$, $\mathbf{S}_2$, $\mathbf{S}_3$ and $\mathbf{S}_4$ of corresponding dimension $s \times s$, $s \times (M - s)$, $(N - s) \times s$ and $(N - s) \times (M - s)$. The matrix $\mathbf{S}_1$ is square and assumed to have an inverse. The existence of this inverse is critical for obtaining derived sensitivities, and if linear dependence occurs amongst the first s concentrations or parameters, a different choice for these quantities must be made. By taking the inverse of $\mathbf{S}_1$ in (51), it is possible to interchange $d\alpha_1 \cdots d\alpha_s$ for $dc_1 \cdots dc_s$ to produce a *derived* equation analogous to (51)

$$(52) \quad \begin{pmatrix} d\alpha_1 \\ \vdots \\ d\alpha_s \\ dc_{s+1} \\ \vdots \\ dc_N \end{pmatrix} = \begin{pmatrix} \mathbf{S}_1^d & \vdots & \mathbf{S}_2^d \\ \cdots & \vdots & \cdots \\ \mathbf{S}_3^d & \vdots & \mathbf{S}_4^d \end{pmatrix} \begin{pmatrix} dc_1 \\ \vdots \\ dc_s \\ d\alpha_{s+1} \\ \vdots \\ d\alpha_M \end{pmatrix}$$

where

(53)
$$S_1^d = S_1^{-1},$$
$$S_2^d = -S_1^{-1}S_2,$$
$$S_3^d = S_3S_1^{-1},$$
$$S_4^d = S_4 - S_3S_1^{-1}S_2.$$

The derived sensitivity matrix elements have the meaning schematically illustrated in (54)

(54)
$$\begin{pmatrix} S_1^d & S_2^d \\ S_3 & S_4^d \end{pmatrix} \begin{pmatrix} \dfrac{\partial \alpha_i}{\partial c_j} & \dfrac{\partial \alpha_i}{\partial \alpha_l} \\ \dfrac{\partial c_k}{\partial c_j} & \dfrac{\partial c_k}{\partial \alpha_l} \end{pmatrix}.$$

The four classes of derived sensitivity coefficients have their own distinct physical meaning. In particular, the elements of S_1^d would arise when the first s concentrations are treated as "measured" and the first s parameters are to be determined from these measurements. This set of inverse sensitivity coefficients has direct application to the analysis and estimation of errors in parameters determined from experimental data (cf. §2.8 below). The derived sensitivities S_2^d measure the degree to which particular parameters may act in concert and still maintain the first s concentrations as fixed observations. These gradients therefore give a measure of model uncertainty in relationship to a chosen body of observations. The elements in S_3^d measure the degree to which potential new measurements are related to the chosen set of s "performed" measurements. These latter gradients could be of utility in considering experimental design or perhaps queuing a set of measurements in their order of significance with regard to observation independence. Finally, elements of S_4^d have the same structure as elementary sensitivity coefficients except now they are constrained under the condition that a subset of concentrations is maintained as fixed. Since the choice of s is arbitrary, a wide variety of derived sensitivities may be calculated.

Second and higher order analogs of the derived sensitivities discussed above may also be calculated. A systematic procedure can be established for this purpose [16]. In addition, analogous procedures may be extended to the functional derivative domain where derived functional coefficients would be involved. In the latter case, the algebraic manipulations connecting (52) and (51) would then become integral equations. Derived sensitivities received their name through the fact that they are "derived" from the elementary sensitivities in (53). However, since any set of M independent parameters and N independent variables should suffice, then the original set of defined variables does not play a unique role (except from the point of view of the original modelling problem). Indeed, the relations in (53) may be readily inverted to express **S** in terms of $\mathbf{S}^d$. Therefore, the question arises as to whether equations may be found for derived sensitivities without prior knowledge of the original elementary set **S**. In the case of functional derivatives, a linear set of integral equations [10] may be obtained for the derived sensitivities without prior knowledge of **S**. However, for the case of constant parameters, a linear set of differential equations analogous to those for elementary sensitivities may not be obtained. Nevertheless, in the temporal case, it is possible to derive nonlinear differential equations for these derived sensitivities [17]. Very little numerical and mathematical exploration of this issue has occurred at this time.

A number of applications of derived sensitivities have been made to kinetic and combustion problems. For example, Fig. 8 shows two derived sensitivity coefficients of a rate constant k_{48} for the reaction

$$H + O_2 + M \xrightarrow{k_{48}} HO_2 + M.$$

The derived sensitivities in this case were achieved by making CO and O_2 independent variables in a multistep CO oxidation scheme [8], [16]. It is evident from the figure that k_{48} is considerably more sensitive to measurements of O_2 than of CO under the conditions of the problem. Another case shown in Fig. 9 corresponds to a derived sensitivity coefficient associated with the reaction-diffusion system containing chemical terms described by (14) [10]. The derived sensitivity $\delta D_1/\delta k_4$ gives a measure of the coupling between diffusion and kinetic processes. The figure was obtained by ex-

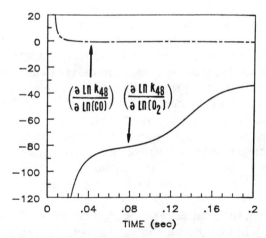

FIG. 8. *Normalized first order derived sensitivity coefficients for the moist oxidation of carbon monoxide $CO/O_2/H_2O$ where k_{48} is the rate constant for the reaction $H + O_2 + M \rightarrow HO_2 + M$. The figure shows that the determination of k_{48} from the measurement of O_2 would require great precision in order to provide a statistically reliable result* [16].

SENSITIVITY ANALYSIS OF COMBUSTION SYSTEMS 81

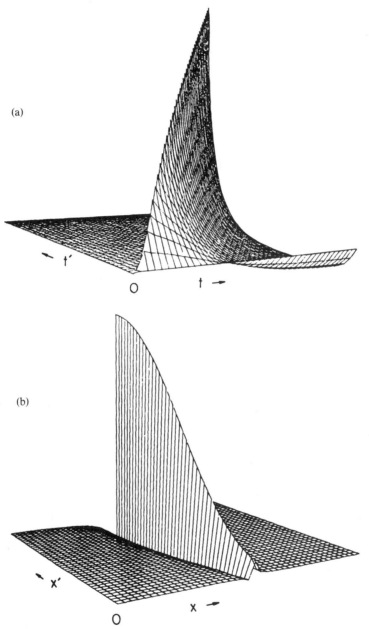

FIG. 9. *The derived sensitivity density* $\delta D_1(x, t; x', t')/\delta k_4$ *corresponding to the reaction-diffusion system with the chemistry in* (14). *Figure* 9(a) *has the spatial coordinates fixed at* $x = 1.0$ *and* $x' = 1.05$ *while Fig.* 9(b) *has the time variables fixed at* $t = 1.05$ *and* $t' = 1.0$. *Strongly localized correlation behavior between the diffusion and rate constant is indicated in both plots.*

changing the chemical species c_1 for the diffusion coefficient D_1. A third example for the use of derived sensitivities in combustion problems has arisen while considering steady state one-dimensional premixed flames [9d]. In these calculations, the flow velocity v in (9) is normally treated as a parameter. Therefore, sensitivity information $\partial c_i/\partial v$ may be obtained in the fashion discussed in §2.2.2 above. However, after performing these calculations, it is physically significant to understand how the flame velocity would be influenced by other system parameters while maintaining the same flame profile. This question is addressed by elements of $\mathbf{S}_1^d$. In particular, calculations of this type were performed by means of the expression

$$(55) \qquad \frac{\partial v}{\partial \alpha_j} = -\frac{(\partial c_i/\partial \alpha_j)}{(\partial c_i/\partial v)}$$

for the model flame in (11) and a realistic hydrogen-oxygen flame.

Derived sensitivities add a considerable new dimension to the utility of sensitivity analysis. A wide variety of questions may be addressed by the coefficients, and the parallel utility of similar manipulations in thermodynamics is worthwhile to remember in this regard.

2.8. Connection to experimental data and error analysis applications of sensitivity information. The role of sensitivity analysis in error estimation is often the first application thought of when the topic is mentioned. Kinetic measurements for the purpose of extracting parameters such as rate constants or transport coefficients will certainly continue to remain important. Reliable values for the extracted parameters will always put demands on experiments performed for such inversion purposes in lieu of the fact that the laboratory data inherently contains errors. An estimation of these latter errors alone does not suffice, since the connection of the measurements to the desired parameters can often be quite complex. This point is clear from considering Fig. 1 where the measurements in the last box are only connected to the input parameters through a complicated inverse kinetics problem. Ideally, it would be best to measure the relevant parameters in "clean" experiments, where the measurements were essentially of the parameters themselves. Unfortunately, this is not possible in many circumstances and the relation between the parameters and measurements is a complex one. Therefore, the canonical problem is one of combining appropriate experimental data with a mathematical model that contains the desired parameters. These parameters would then be fitted to the data and in the process, estimates extracted for the errors in the parameters as a result of experimental uncertainties and model uncertainties.

In order to connect modelling and sensitivity analysis with the data generated in experiments, it is necessary to define an appropriate optimizing function (functional) relating the two types of information. A least squares functional of the type in (39) is commonly chosen and for simplicity here, we will take the data as available at discrete times. In a purely time-dependent problem, the residual to be minimized may be taken to have the form

$$(56) \qquad R = \sum_{i=1}^{n_0} [(c^e(t_i) - c(t_i, \boldsymbol{\alpha}))/c^e(t_i)]^2$$

where weighted least squares is chosen relative to the magnitude of the actual observations. The function $c^e(t_i)$ is the experimentally measured concentration at time t_i and $c(t_i, \boldsymbol{\alpha})$ is the corresponding concentration obtained from the temporal model. Similar expressions and procedures could be considered if spatial measurements were performed. Typically, only a few members of the parameter vector $\boldsymbol{\alpha}$ would actually be varied in order to fit a given set of data. The remaining parameters would therefore be assumed "known" from previous measurements or calculations. The sum in (56) is over the available n_0 time-dependent data points. Differentiation of (56) with respect to the varied parameters along with the requirement $\partial R/\partial \alpha_i = 0$ will produce an equation implying a relationship between the parameters and the observations

$$(57) \qquad \boldsymbol{\alpha} = \boldsymbol{\alpha}(c^e(t_1), c^e(t_2), \cdots, c^e(t_{n_0})).$$

Therefore, differentiation once again with respect to an element of the set of observations will ultimately yield an equation for the sensitivity of the model parameters to the experimental data $\partial \ln \alpha_i/\partial \ln c^e(t_n)$ where logarithmic normalization has been chosen,

$$(58a) \qquad \sum_{k=1}^{n_\alpha} Q_{jk}(\partial \ln \alpha_k/\partial \ln c^e(t_h)) = -L_{jh}$$

with n_α being the number of parameters to be determined and

$$(58b) \qquad Q_{jk} = \sum_{i=1}^{n_0} \left(\frac{\partial^2 \ln c(t_i)}{\partial \ln \alpha_j \partial \ln \alpha_k}\right)\left(\frac{c(t_i)}{c^e(t_i)}\right)\left(1 - \frac{c(t_i)}{c^e(t_i)}\right)$$
$$+ \left(1 - \frac{2c(t_i)}{c^e(t_i)}\right)\left(\frac{c(t_i)}{c^e(t_i)}\right)\left(\frac{\partial \ln c(t_i)}{\partial \ln \alpha_j}\right)\left(\frac{\partial \ln c(t_i)}{\partial \ln \alpha_k}\right),$$

$$(58c) \qquad L_{ji} = \left(1 - \frac{2c(t_i)}{c^e(t_i)}\right)\left(\frac{c(t_i)}{c^e(t_i)}\right)\left(\frac{\partial \ln c(t_i)}{\partial \ln \alpha_j}\right).$$

In order to solve (58a), it is necessary to have available both first and second order sensitivity coefficients in (58b) and (58c). The second order coefficients occur to take into account the inevitable differences arising between the model and experimental results. The solution of (58a) will yield the necessary coefficients to estimate the covariant matrix of the determined parameters.

Standard procedures from statistics will yield the following expression for the parameter covariance matrix:

$$\text{cov}(\Delta\alpha_j/\alpha_j, \Delta\alpha_k/\alpha_k)$$

(59)
$$= \sum_{i=1}^{no} \sum_{h=1}^{no} (\mathbf{Q}^{-1}\mathbf{L})_{ji}(\mathbf{Q}^{-1}\mathbf{L})_{kh} \text{cov}(\Delta c^e(t_i)/c^e(t_i), \Delta c^e(t_h)/c^e(t_h))$$
$$+ \sum_{l}' \sum_{l}' (\mathbf{S}_2^d)_{jl}(\mathbf{S}_2^d)_{km} \text{cov}(\Delta\alpha_l/\alpha_l, \Delta\alpha_m/\alpha_m)$$

where the last term concerns errors in the assumed "known" model parameters and the prime on the sum indicates that they do not include parameters on the left-hand side of the equation. In most problems, the correlation between different sets of experimental data is unknown, and therefore the data covariance matrix on the right-hand side of (59) is reduced to being diagonal. When utilizing (59), it is important to understand that this error estimate corresponds to only the lowest order term in a Taylor series relating parameter uncertainties to experimental errors and other modelling errors. Nevertheless, the estimates obtained from (59) can be an extremely helpful guide. As an example in the case of combustion, this procedure was recently applied to extracting the rate constant for the reaction [18]

$$H_2CO + OH \to H_2O + HCO$$

where the data was obtained from flow tube measurements.

A few final comments are worthy concerning the above analysis and its relation to other aspects of modelling and sensitivity theory. First note that in the case where the experimental data and modelling calculations coincide to some acceptable precision, $c^e(t_i) = c(t_i)$, then it may be shown that the solution to (58) reduces to

(60)
$$\frac{\partial \ln \alpha_k}{\partial \ln c^e(t_h)} \simeq (\mathbf{S}_1^d)_{kh} = (\mathbf{S}_1^{-1})_{kh}$$

where $\mathbf{S}_1^d$ is the derived sensitivity matrix referred to in (53) now made up of log normalized derivatives. The estimate in (60) for the coefficients $(\mathbf{Q}^{-1}\mathbf{L})_{kh}$ in (59) is the one most often used in practice. However, serious discrepancies can arise in cases having significant differences between the model calculations and experimental data. Another point is that statistical errors could arise in the model due to the computational procedures and an expression exactly analogous to the first term of (59) may be produced with $\mathbf{L}$ replaced by the matrix $\mathbf{M}$ having elements

(61)
$$M_{ji} = [c(t_i)/c^e(t_i)]^2 \frac{\partial \ln c(t_i)}{\partial \ln \alpha_j}.$$

The resultant term added to (59) would require an estimate for the statistical uncertainties in the computational results. A third topic involves the non-statistical contribution of model uncertainties to parameter estimates from experimental data. In order to obtain this information, it is necessary to invert the sensitivity relation $d\mathbf{c} = \mathbf{S}d\boldsymbol{\alpha}$. The sensitivity matrix $\mathbf{S}$ is typically not square and a generalized inverse must be sought. A least squares estimator can be used to give rise to the following expression for $\Delta\alpha_k/\alpha_k$ taken to first order,

$$(62) \qquad \Delta\alpha_k/\alpha_k = \sum_{i=1}^{no} \{(S^TS)^{-1}S^T\}_{ki}\Delta c(t_i)/c(t_i).$$

The practical utility of this equation requires an estimate for the model uncertainty on the right-hand side. An extremum estimate of this could be achieved by assuming the entire discrepancy between the experimental and model results is due totally to the model $\Delta c(t_i)/c(t_i) \simeq [c(t_i) - c^e(t_i)]/c(t_i)$. A final point to emphasize concerning error analyses is that sensitivity techniques may be immediately applied to estimate predicted observable uncertainties from corresponding parameter uncertainties in a model

$$(63) \qquad \begin{aligned} &\text{cov}(\Delta c_j(t)/c_j(t), \Delta c_k(t)/c_k(t)) \\ &= \sum_{i=1}^{M}\sum_{h=1}^{M} \left(\frac{\partial \ln c_j(t)}{\partial \ln \alpha_i}\right)\left(\frac{\partial \ln c_k(t)}{\partial \ln \alpha_h}\right) \text{cov}(\Delta\alpha_i/\alpha_i, \Delta\alpha_h/\alpha_h). \end{aligned}$$

Only the first order elementary sensitivity coefficients are required to provide these estimates. Modelling would perhaps be treated as a more respected tool if (63) was routinely applied.

2.9. Sensitivity to missing model components. A criticism commonly raised when considering the predictions of a model concerns the issue of how does one know that a given model is actually complete. In practice, of course, the answer to this question will remain strictly unknown. However, in realistic problems, a partial answer can often be obtained. In particular, modelling is usually begun by assembling a lengthy list of potential physical processes and their accompanying mathematical expressions or equations. Even in the simplest problems, this list can be far too long for full implementation in a computer model. Inevitably, a reduced model is chosen for testing and evaluation. Assuming the results of the model show discrepancies with available experimental data and assuming these discrepancies are due to the model, it is possible to make an estimate of the uncertainty associated with the missing model component.

In order to carry out the estimate indicated above, two criteria must be satisfied. First, a list of *possible* candidate reactions, transport processes or other physical factors must be available. Second, the dependent variables associated with these additional processes must already exist in the original model (e.g., a reaction requiring or producing a new species would not be allowed). Assuming these criteria are satisfied, it is an easy matter to see how sensitivity coefficients with respect to the missing parameters could be obtained. In a sense, the original reduced model contains these parameters except they are evaluated at a zero nominal value. Despite the fact the parameters are set to zero, this does not imply that their gradients will be null. Therefore, an immediate application of elementary sensitivity analysis can be made with respect to the missing model parameter by employing the various standard procedures for generating sensitivity coefficients in §2.B above. For example, in the purely temporal case, (7) could immediately be applied where the term $\partial R/\partial \alpha_j$ would be evaluated at the nominal value of all the system parameters including the new null ones. This procedure has been applied in an extensive model involving the pyrolysis of propane [19].

3. New directions and issues related to sensitivity analysis. The introduction indicated that there were a number of topics related to sensitivity analysis and this section will discuss some of these items. In addition, there are questions and directions for study worthy of exploration which have not been treated thus far. Since each of the topics in this section are only loosely connected, they will each be separately discussed in a format similar to that of §2.

3.1. Choice of system parameters. Throughout the analysis in §2, the original parameters chosen in (1) and (2) were assumed to be those of interest for probing by various sensitivity techniques. Although presumably these parameters entered in a natural physical way into these equations, there may be circumstances when a different choice would be better. A hint of this issue was evident in §2.6 with the extreme case of the output being characterized by its own parameters $\beta_1, \beta_2, \cdots$. Unfortunately, the relationship between these parameters and the original ones (cf. (38)) is generally unknown. Indeed, the purpose of feature sensitivity analysis was to provide information about this relationship. This perspective can be extremely valuable for experimental considerations, and it may also be expressed in a somewhat more flexible fashion. In particular, the question arises as to whether a judicious choice of parameters

(64) $$\gamma = \gamma(\alpha)$$

can be made prior to solving the system of differential equations. There is surely an infinite number of mathematically acceptable linear and nonlinear transformations of this type. Making a reasonable choice will require appropriate physical and mathematical guidance. For example, it may be desirable to identify a set of parameters $\gamma_1, \gamma_2, \cdots$ such that a subset of the chemical species are invariant $\partial c_i/\partial \gamma_j = 0$ or at least weakly dependent upon these variables. Such a transformation would be tantamount to finding the level curves in some region of the original parameter space associated with the concentrations of interest. The full solution to this problem is equivalent to achieving global parameter space mapping discussed in §3.6 below. Less ambitious goals may be appropriate in many cases and taking the transformation (64) to lowest order (or assuming it is linear), we will arrive at the new sensitivity coefficients

(65) $$\left(\frac{\partial c_i}{\partial \gamma_j}\right) = \sum_l \left(\frac{\partial c_i}{\partial \alpha_l}\right)\left(\frac{\partial \alpha_l}{\partial \gamma_j}\right)$$

which expresses the new sensitivity coefficients as a linear transformation of the traditional set. The flexibility in choosing the input parameters should be given consideration at the initial stages of modelling and sensitivity analysis; proper choices can simplify the ensuing physical interpretation and increase the utility of sensitivity techniques for applications such as in control problems.

3.2. Computational aspects of sensitivity analysis. Various parts of §2 referred to the available numerical techniques for obtaining sensitivity coefficients. This topic is one of rapid development and historically it has been a serious bottleneck for the implementation of sensitivity techniques. In addressing the issue, the first problem of simply solving the original model equations must be considered. Great effort has gone into more efficiently solving the modelling equations, particularly when the equations are nonlinear. In this sense, relatively little effort has been put forth for the sensitivity equations. Mathematically a special characteristic of the sensitivity equations is their inherent linear form. This aspect forms the basis of some of the numerical techniques. Currently the approach that seems most promising comes from the recognition that the sensitivity equations are driven by the same Jacobian matrix often utilized for implicitly solving the original modelling equation [7], [9]. It is anticipated that current coding will attempt to fully exploit this similarity wherever possible. However, even with this development it is still not entirely clear whether the resultant algorithms will actually be optimal. Given the broad physical utility of sensitivity analysis in modelling, it is suggested that code development take this goal into account at the earliest stages. In particular, it would be desirable to consider the modelling *and* sensitivity equations together as the goal for solution and develop codes on this basis. The present efforts represent a step in this direction, but further improvement would certainly be welcome.

An issue related to the above points concerns using sensitivity analysis as a guide to the numerical solution of the modelling equations. Regardless of the detailed aspects of the numerical procedure, solution of the model will inevitably involve some type of discretization (e.g., collocation, finite differences, basis set expansions, etc.) of the differential equations. A natural question then arises concerning the accuracy and stability of the numerical procedure with regard to the discretization. For example, supposing that a finite difference technique is utilized for a steady state problem, it would be interesting to know the degree to which the solution at one point is sensitive to the mesh locations throughout the domain. This question would be addressed by the nonlocal type of sensitivity coefficient $\partial c_i(x_l)/\partial x_n$ where x_l and x_n are particular mesh points [20]. One might anticipate that physically and mathematically stable problems would have the characteristic that the solution at point x_l would become increasingly less sensitive to meshing at points x_n far away. Various types of combustion phenomena can show unstable behavior which may, in turn, exhibit itself in highly nonlocal meshing sensitivity. The calculation of these nonlocal mesh sensitivity coefficients *during* the solution to the modelling equations could provide a useful guide for more efficient coding, especially in adaptive gridding procedures where points are to be added or relocated.

3.3. Hierarchical modelling and sensitivity analysis. The physical focus of this chapter has been on combustion phenomena and in a broader sense kinetics and transport processes. Equations of the type in (1) describing these phenomena are often referred to as phenomenological for dealing with bulk properties. The physical description produced by these models has been shown to be accurate provided the models themselves are physically correct. Nevertheless, combustion processes have roots going back to the fundamental events occurring at the atomic scale. In recognition of this point, there have been numerous efforts [21] to start with the fundamental coulombic interactions between the nuclei and the electrons and proceed forward through Schrödinger's equation to ultimately calculate rate constants for input to (1). Carrying out these calculations for realistic reactions can be an arduous task often requiring approximations at several levels. As computational techniques improve, fewer of these approximations will be needed and this *ab initio* approach will surely take on increased importance as a source of information for kinetic modelling.

The sensitivity techniques discussed in this paper have already been applied to quantum dynamics at a microscopic atomic scale. However, little effort has been given to proceeding through the hierarchy leading from the microscale to macroscopic observables. The steps in this process are schematically illustrated in Fig. 10. Although there are no serious questions about the nature of the nonrelativistic interactions of the electrons and nuclei, sensitivity concerns could arise regarding the numerical solution of the electronic structure problem leading to a potential energy surface. Leaving this issue aside, a realistic starting point is the molecular Hamiltonian and particularly the intermolecular potential portion in the Hamiltonian. Sensitivity techniques have been utilized to probe the relationship between collision

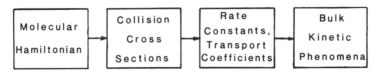

FIG. 10. *The hierarchical flow starting from fundamental interactions and proceeding through to bulk kinetic observables. A diagram like that of Fig. 1 could be written for each step in the present plot. This chapter has primarily focused on the last step connecting kinetic or transport coefficients to bulk phenomena. Sensitivity analysis may be carried out at all levels ultimately addressing the degree to which microscopic information is important at the macroscopic level.*

cross sections and various features or parameters in the molecular Hamiltonian, although applications have thus far not been made to reactive processes [5]. Steps further along the hierarchy either involve additional integrations/summations over appropriate variables or the solution of kinetic-transport equations. Therefore, it seems natural to expect, for example, that the chemical concentration or temperature profiles in a combustion reactor are probably highly insensitive to details of the intermolecular potential at the atomic scale. This conclusion is surely not entirely correct since there would be no variability, for example, between the performance of various fuels. Therefore, the basic question remains as to what characteristics at the atomic scale are relevant to macroscopic considerations? There may also be cases of bulk phenomena which are highly sensitive to certain details at the atomic scale. The understanding of these problems is of fundamental concern and could have significant impact on the theoretical and computational procedures for calculating kinetic constants as well as on the design of experiments for inversion of laboratory data back to microscopic level information. Sensitivity analysis should be able to make a clear contribution to these questions by probing the degree to which information is lost or even magnified while proceeding through the hierarchy in Fig. 10. In addition, various derived sensitivity coefficients should be valuable for giving insight into inversion issues. Many of the available sensitivity and modelling tools are capable of handling various aspects of these problems.

3.4. Network theory and sensitivity analysis. The equations of chemical kinetics are typically of a coupled nature representing the chemical reactions taking place. When attempting to understand the solutions to these equations, it is common practice to consider the connectivity or flow of reactions. This entangled coupling is largely responsible for the complexity of the relationship between the input and output in Fig. 1. However, recent efforts have shown that much can be learned from the topological or network structure of the coupled differential equations [22]. For example, a primary concern in many systems is the stability of the kinetic trajectories (cf. §2.5) or perhaps whether the system will go to a fixed steady state or a limit cycle. Rather surprisingly, considerable progress has been made on answering these questions by directly examining the network structure of the differential equations *without* actually solving them. Such an analysis can be extremely valuable, but it will of course not predict the actual trajectory or orbit in state space. In general, network structure alone will also not likely be able to describe the detailed role of a given parameter. Therefore, these techniques represent an interesting complementary procedure to sensitivity analysis which tends to be more detailed and local in its perspective. There may also be a possible connection between the procedures by seeking the sensitivity of the solution with respect to the network structure rather than merely to individual rate constants or other parameters. Clearly techniques beyond those discussed in §2 will be needed to address this problem and its significance certainly warrants further study.

3.5. Lumped system reduction and sensitivity analysis. The general purpose of mathematical modelling is to produce models that are physically correct and practically viable for making predictions and controlling devices of concern. In particular, this goal has been a longstanding one in the case of engineering combustor design and control. There are two broad approaches to developing models for this purpose. First, one may start with a rather complete list of physical and chemical processes, but only select a very minimal model from this set for testing. Typically, such a model will fall short of its objectives and a more sophisticated model will then be called for. Under these conditions, sensitivity analysis of missing model components discussed in §2 could be applied. The other procedure would consist of running the model with all possible physical factors included. Assuming the parameter values are correct, such a model would presumably give proper results. However, a comprehensive model of this type would not be of practical utility for routine application in the control and design of engineering devices. In addition, when only certain performance characteristics are of concern for engineering purposes, then all the details of the comprehensive models are likely not necessary. Therefore, the second approach would attempt to achieve a lumped dependent and/or independent variable model which is adequate for the questions of concern.

Lumping efforts have been carried out in a number of areas including industrial applications [23]. Thus far, these efforts have largely been guided by information gained from the tedious task of running the models many times under different parameter conditions. Sensitivity techniques would appear to have the capability of providing systematic tools for quantitatively guiding lumping efforts. For example, the elementary sensitivity coefficients often give an immediate measure of which parameters act in a similar fashion or which dependent variables have similar sensitivities. Therefore, lumping may be suggested at the elementary sensitivity level. In addition, similar coordinate and/or time behavior of elements of the Green's function matrix provides another measure of which species act in concert. Furthermore, an eigenvalue-eigenvector analysis of the Green's function could be utilized to deduce which dependent variables are responsible for long-term or transient behavior, depending on the time scale of interest for the particular application. Another tool of possible utility would be rank reduction (within a specified tolerance) of various sensitivity matrices. A reduced rank would indicate that only a subclass of variables actually act independently of each other. Once lumped parameters and dependent variables have been achieved, an interesting question concerns the relationship of these variables to those in the more elaborate detailed model. The techniques in §2.6 are applicable to addressing this latter question. Lumping is an advanced realm of modelling and sensitivity analysis with significant potential practical applications.

3.6. Global parameter space mapping techniques. All the techniques of this paper have thus far been confined to what are referred to as local procedures. This labelling arises due to the fact that gradient sensitivity analysis explores the local vicinity of an operating point α^0 in parameter space. A great deal of physical information can be extracted by exploring the vicinity of the operating point, but there still remains the ultimate issue of the response to finite excursions throughout parameter space. Full knowledge of the system and its behavior in the vicinity of the operating point may in fact give scant information about how the system would behave at a finite distance away from the operating point. Understanding of this issue would constitute a global parameter space map. A full achievement of this map in many respects represents an *inherently* difficult problem which will continue to remain computationally difficult. Nevertheless, it is an important problem particularly for design and control issues in engineering and a host of scientific problems. The traditional approach of Monte Carlo sampling is simply out of the question for any realistic multiparameter systems. A Lie algebraic approach does seem to be feasible for treating this problem in a more practical vein and an outline of the technique as it is now understood is given below [24].

Within the framework of traditional sensitivity analysis, a finite excursion away from the operating point would be treated by the use of a Taylor series with the sensitivity coefficients

(66)
$$c_i(\alpha^0 + \Delta\alpha) = c_i(\alpha^0) + \sum_j \left(\frac{\partial c_i}{\partial \alpha_j}\right)_{\alpha^0} \Delta\alpha_j \\ + \tfrac{1}{2}\sum_{jk} \left(\frac{\partial^2 c_i}{\partial \alpha_j \partial \alpha_k}\right)_{\alpha^0} \Delta\alpha_j \Delta\alpha_k + \cdots.$$

For reasons of computational practicality, typically terms only up to second order are available for the series and hence sophisticated series extension techniques such as Padé approximants are not approximate. In order to understand the basic aspects of the Lie approach, it is first necessary to extend the space of concern to include all of the parameters, coordinates and time, as well as the system dependent variables. As a specific case, consider the temporal system described in (3). A first order differential operator referred to as the group generator U may be defined having the following form

$$(67) \qquad U = \sum_l \psi_l \frac{\partial}{\partial \alpha_l} + \sum_{l'} \xi_{l'} \frac{\partial}{\partial c_{l'}}$$

where the space of variables has been restricted to the parameters and concentrations. The coefficients ψ and ξ are generally functions of the parameters, species and time. The role of the generator may be understood by noting that an infinitesimal excursion ($\epsilon \ll 1$) through parameter space may be achieved by $1 + \epsilon U$ acting in the hyperspace defined above. In general, a given problem may have a set of group generators corresponding to all possible independent excursions through the space. The unknown functions ψ and ξ in (67) are determined by demanding that $1 + \epsilon U$ leave (3) invariant to its operation. In other words, the action of $1 + \epsilon U$ on (3) would preserve the *form* of the equation although all of the parameters and variables may be altered by the action of the generator. The demand that (3) be left invariant by the action of the differential operator in (67) will produce a set of differential equations for the unknown coefficients of the form

$$(68) \qquad \frac{\partial \xi_{l'}}{\partial t} + \sum_i \frac{\partial \xi_{l'}}{\partial c_i} R_i - \sum_i \xi_i \frac{\partial R_{l'}}{\partial c_i} - \sum_l \psi_l \frac{\partial R_{l'}}{\partial c_l} = 0.$$

Equation (68) is to be understood with both time and concentrations all treated as independent variables. It is intriguing to observe that these equations encompass those of traditional sensitivity analysis in (4a) by specification of a particular trajectory for analysis. This observation clearly indicates the broader generality of the Lie approach. The practical (or at least approximate) solution of (68) is critical to the practical implementation of this technique and it is too early to tell what procedures will emerge. By following logic similar to that of transformations in quantum mechanics, one may show that finite excursions will be generated by the operator

$$(69) \qquad T = \exp(\epsilon U).$$

The variable ϵ is said to be a group parameter chosen according to the magnitude of the desired excursion.

In essence, the transformation in (69) will describe the allowed trajectories through the hyperdimensional space that leave the equations of motion invariant. The significance of this latter issue may be understood in regard to the use of traditional sensitivity analysis alone. The latter technique may provide infinitesimal excursions in parameter space although it provides no guarantee that the new "solution" will solve an equation of exactly the same form as the original system. This distinction is at the heart of the difference between sensitivity analysis and the Lie techniques. It is important to preserve the form of the original equations, since, by definition, the equations are valid at all accessible points in parameter space.

A number of issues need to be explored, including methods for performing the exponentiation in (69), in order for practical results to be achieved from the Lie technique. This topic is certainly worthy of study since a variety of problems would be readily opened up by the method. Applications to control theory and system design would be especially amenable to study. For example, it would be desirable to delineate the region of control space which would preserve the stability of a combustion system. This example is only one of a variety of questions concerning constrained excursions through parameter space. The constraints may be built directly into the generators U which is not possible by the application of traditional Monte Carlo sampling procedures.

4. Conclusions. It was the intention of this chapter to give an overview of sensitivity techniques, as they are currently available, with particular emphasis towards problems in the combustion domain. Although sensitivity analysis as a formal subject has existed for many years, it is only relatively recently that the subject has blossomed and particularly come into the combustion and kinetics fields. Some aspects of sensitivity analysis are already at the applied stage while in other areas further basic developmental work is needed. Due to the valuable information available through sensitivity analysis, it is hoped that the techniques eventually become a routine part of mathematical modelling efforts.

REFERENCES

[1] R. TOMOVIC AND M. VUKOBRATOVIC, *General Sensitivity Theory*, American Elsevier, New York, 1972.
[2] P. M. FRANK, *Introduction to System Sensitivity Theory*, Academic Press, New York, 1978.
[3] J. TILDEN, V. COSTANZA, G. MCRAE AND J. SEINFELD, in Modelling of Chemical Reaction Systems, K. Ebert, P. Deuflhard and W. Jager, eds., Springer-Verlag, Berlin, 1981.
[4] D. COX AND P. BAYBUTT, *Methods for uncertainty analysis: A comparative survey*, Risk Analysis, 1 (1981), pp. 251–258.
[5a] H. RABITZ, *Chemical sensitivity analysis theory with applications to molecular dynamics and kinetics*, Comput. Chem., 5 (1981), pp. 167–180.
[5b] L. ENO AND H. RABITZ, *Sensitivity analysis and its role in quantum scattering theory*, Adv. Chem. Phys., 51 (1982), pp. 177–226.

[6] H. RABITZ, M. KRAMER AND D. DACOL, *Sensitivity analysis in chemical kinetics*, Ann. Rev. Phys. Chem., 34 (1983), pp. 419–461.
[7] B. LOJEK, *Sensitivity analysis of nonlinear circuits*, IEEE Proc., 129G (1982), pp. 85–94.
[8] R. YETTER, F. DRYER AND H. RABITZ, *Some interpretive aspects of elementary sensitivity gradients in combustion kinetics modelling*, Combust. and Flame, 59 (1985), pp. 107–133.
[9a] W. STEWART AND J. SORENSEN, in Chemical Reactor Engineering, Proc. VIth European Symposium, (1976), pp. 1-12.
[9b] C. IRWIN AND T. O'BRIEN, DOE/METC Tech. Rep., 53, Morgantown, W.V.
[9c] A. DUNKER, Atmospheric Environment, 15 (1981), pp. 1155ff.
[9d] Y. REUVEN, M. SMOOKE AND H. RABITZ, *Sensitivity analysis of boundary value problems: Application to nonlinear reaction-diffusion systems*, to be published.
[10] R. LARTER, H. RABITZ AND M. KOBAYASHI, *Derived sensitivity densities in chemical kinetics: A new computational approach with applications*, J. Chem. Phys, 79 (1983), pp. 692–707.
[11a] S. GROSSMANN, *Langevin forces in chemically reacting multicomponent fluids*, J. Chem. Phys., 65 (1976), pp. 2007–2012.
[11b] J. KEIZER, *Concentration fluctuations in chemical reactions*, J. Chem. Phys., 63 (1975), pp. 5037–5043.
[11c] D. DACOL AND H. RABITZ, *Sensitivity analysis of stochastic kinetic models*, J. Math. Phys., 25 (1984), pp. 2716–2727.
[12] W. HORSTHEMKE AND R. LEFEVER, *Noise Induced Transitions*, Springer-Verlag, New York, 1983.
[13] D. DACOL AND H. RABITZ, unpublished work.
[14] R. HEDGES AND H. RABITZ, *Stability and sensitivity analysis of initial value problems with applications to kinetics and classical mechanics*, J. Chem. Phys., in press.
[15a] D. CACUCI, *Sensitivity theory for nonlinear systems. I. Nonlinear functional analysis approach*, J. Math. Phys., 22 (1981), pp. 2794–2812.
[15b] J. G. B. BEUMÉE, L. ENO AND H. RABITZ, *Pointwise feature sensitivity analysis*, J. Comp. Phys., 57 (1985), pp. 318–325.
[16] R. YETTER, F. DRYER, L. ESLAVA AND H. RABITZ, *Elementary and derived sensitivity information in chemical kinetics*, J. Phys. Chem., 88 (1984), pp. 1497–1507.
[17] B. GUZMAN AND H. RABITZ, *On the relationship between derived sensitivities and elementary sensitivities in temporal kinetics*, to be published.
[18] R. YETTER, B. KLEMM, F. DRYER AND H. RABITZ, to be published.
[19] D. EDELSON AND D. ALLARA, *A computational analysis of the alkane pyrolysis mechanism. Sensitivity analysis of individual reaction steps*, Int. J. Chem. Kinet., 12 (1980), pp. 605–621.
[20] P. PEDERSON, *Some properties of linear strain triangles and optimal finite element models*, Int. J. Num. Meth. Engrg., 7 (1973), pp. 415–429.
[21] W. MILLER, *Dynamics of Molecular Collisions*, Plenum Press, New York, 1976.
[22a] M. FEINBERG, in Dynamics and Modelling of Reactive Systems, W. Stewart, W. Ray and C. Conley, eds., Academic Press, New York, 1980.
[22b] B. CLARKE, *Stability of complex reaction networks*, Adv. Chem. Phys., 43 (1980), pp. 1–215.
[22c] R. LARTER AND B. CLARKE, *Chemical reaction network sensitivity analysis*, Proc. Workshop on Fluctuations and Sensitivity in Non-Equilibrium Systems, Austin, TX, 1984, to appear.

[23] D. NACE, S. VOLTZ AND V. WEEKMAN, *Application of a kinetic model for catalytic cracking*, Ind. Eng. Chem., 10 (1971), pp. 530–537.
[24a] C. WULFMAN AND H. RABITZ, *A Lie algebraic approach to parameter space mapping for systems described by ordinary differential equations*, to be published.
[24b] G. WANG, T. TARN AND J. CLARKE, *On the controllability of quantum-mechanical systems*, J. Math. Phys., 24 (1983), pp. 2608–2618.

CHAPTER III

Turbulent Combustion

F. A. WILLIAMS

1. Introduction. Like the proverbial elephant, turbulent combustion exhibits so many different aspects that two investigators, chosen at random from the set of all investigators examining the subject, will with high probability disagree entirely about its general characteristics. For example, few will know the full definitions of all of the terms "k-ϵ modelling," "strange attractor," "Favre average," "conserved scalar," "Kolmogorov scale," "functional differentiation," "coherent structure," "large-eddy Damköhler number," "random vortex," "spectral transfer," "intermittency" and "age theory." Almost all of the investigators will consider at least one of the items in this sequence to be practically irrelevant to turbulent combustion. Yet, for each of the items there likely exist investigators who would identify that item as the central aspect of the subject. Turbulent combustion has now become so multifaceted that no one has the background necessary to give a complete and impartial review of it. Therefore the reader must prepare himself for an incomplete and somewhat biased review.

Fifteen years ago, in the introduction to a review on laminar combustion [1], it was remarked that there had been little progress in the theory of turbulent combustion, and it was predicted that the future rate of progress would not be much greater than that of the immediate past. The truth of that prediction was short-lived (just as with most predictions in turbulent combustion). Research in the subject has mushroomed in numerous diverse directions in recent years, and a few of the avenues have led fruitfully to genuine discoveries of lasting value. The intent of the present article is to attempt to highlight the most promising avenues and to identify some of the pitfalls along the many seductively ruinous paths.

Researchers have been attracted to turbulent combustion for two very different reasons. On the practical side, it is well recognized that there is turbulence in the great majority of combustion devices. Therefore whoever is interested in worldly problems may be motivated to study the subject. On the fundamental side, the complexities of processes of turbulent combustion make them most challenging topics of investigation—an average graduate student cannot be expected to produce a significant fundamental contribution. Therefore anyone who is seeking challenges and is captivated by the intricacies that can be seen at the hearth of an open fire runs the risk of being drawn toward analysis of the subject.

A large number of reviews demonstrate that indeed many have been drawn to turbulent combustion. Over fifty reviews can be cited, but only a few of the recent ones are referenced here [2]–[11]. The diversity of viewpoints would be at least partially appreciated by the reader if he were to compare these reviews. One of the sources for more than one viewpoint is a 250-page book on the subject [2].

In the following presentation some of the viewpoints and methods that have been used in analysis of turbulent combustion will be indicated first. Next, an identification of regimes of turbulent combustion will be offered, with an exhortation that approaches be associated explicitly with particular regimes. The subject is so complicated that it would be extremely unlikely to find one method equally useful in all regimes. Later sections are devoted to assessments of what is known or might soon be discovered in various regimes.

2. Analytical approaches. Prior to discussion of approaches to analysis, agreement must be reached on just what turbulent combustion is. There is not a universal acceptance of the conservation equations that describe turbulent combustion. Are equations of fluid flow an appropriate beginning? Is it necessary to proceed instead to a kinetic-theory or quantum-mechanical level of description? Arguments in favor of the latter positions can be offered. For example, kinetic-theory studies have found particular long-range correlations that modify chemical reaction-rate expressions from those of usual fluid equations for certain problems in turbulent combustion [12]. One thing that is clear is that if it is necessary to go to a deeper level of description than fluid equations, then the potential complexity of the subject is increased. Therefore let it be agreed that turbulent combustion is described by the Navier–Stokes equations, suitably augmented by chemical-kinetic laws. Assuredly much of the reality of turbulent combustion remains present at this level.

At the fluid level of description three-dimensional, time-dependent differential equations for conservation of mass, momentum, energy and chemical species apply to turbulent combustion (e.g., Chapter I, §§1, 2). Choices still remain concerning the nature of the boundary and initial conditions for these equations. Uncertainties in these choices arise equally for chemically reacting and nonreacting flows. Therefore they can be discussed without reference to chemical reactions.

Turbulence usually is considered to be a stochastic phenomenon; at least it always possesses probabilistic aspects. A conceptually simple way to introduce stochastics is to let the boundary and/or initial conditions be stochastic. There is a prevalent belief, supported by experiment, that if this is done, then the statistical properties of the turbulent flow, throughout the domain of real interest, exhibit no dependence on the statistical aspects that have been introduced at the boundaries. The statistics of the turbulence is dominated by the many fluid instabilities that occur in turbulent-flow experiments. Even if a proper set of deterministic boundary and initial conditions is imposed on the fluid equations (so that, according to common belief, the flow field is determined uniquely) under conditions termed turbulent, the flow is thought to become chaotic or pseudorandom in that it passes tests for randomness. A realization of a turbulent flow may be highly sensitive to its specific boundary and initial conditions while the statistical properties are insensitive thereto. It is in this sense that turbulence possesses a strange attractor, an entity about which a number of popular accounts have been published recently (e.g., [13]). Ideas related to strange attractors are too new for their potential impact on concepts of turbulent combustion to be assessed properly.

2.1. Evolution of probability-density functions. Once turbulence is accepted as a stochastic process, probabilistic approaches to its description become appropriate. In one of these approaches, attention is focused on the evolution of probability-density functions [10]. As is demonstrated, for example, in E. E. O'Brien's contribution to [2], partial differential equations may be derived for the evolution of probability-density functions for the velocity and state variables of turbulent combustion. The character of the approach is traced to earlier work on nonreacting Navier–Stokes turbulence [14], [15].

For reacting gas mixtures having N different chemical species there are $N + 4$ random functions of space and time, which may be taken to be the mass per unit volume for each species, temperature and the three components of velocity. A probability-density functional $P[V(x, t)]$ for these functions may be defined so that the probability of finding them in the range $\delta V(x, t)$ about the $(N + 4)$-vector function $V(x, t)$ is $P[V(x, t)]\delta V(x, t)$, where $\delta V(x, t)$ is a volume element in the function space. Thus, $\delta V(x, t)$ denotes a narrow range of functions that includes a continuously infinite number of functions, all of those remaining wholly within the dark region of Fig. 1. A linear functional differential equation for $P[V(x, t)]$ is readily derived from the conservation equations. This equation is complete in that in principle it contains all of the statistical information about the flow.

Since the functional differential equation cannot really be solved, probabilistic descriptions at less detailed levels of information must be sought. For example, with respect to the statistics, attention may be focused on particular values of x and t, so that the random functions become random

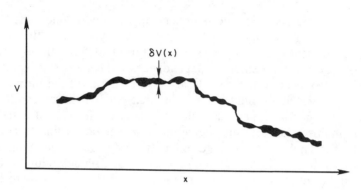

FIG. 1. *Illustration of volume element in function space.*

variables [16], and at the simplest level probability-density functions $P(v)$ may be sought, such that the probability that a random variable lies in the range dv about the value v is $P(v)dv$. At intermediate levels of description joint probability-density functions for, say, M random variables are sought so that, for an M-vector **v** with components $v_1, v_2, \cdots, v_M$, the generalization of $P(v)dv$ is $\mathbf{P}(\mathbf{v})dv_1 dv_2 \cdots dv_M$. No matter how large M is, so long as it is finite any set of equations derived formally for any finite collection of functions $\mathbf{P}(\mathbf{v})$, e.g., by performing suitable integrations of the functional differential equation for $\mathbf{P}[\mathbf{V}(\mathbf{x}, t)]$, contains more unknowns than there are equations. This inconvenient state of affairs is an illustration (here at the most fundamental level) of the infamous *closure* problem of turbulence that pervades nearly all approaches to the subject.

Closure is the Achilles heel of theories of turbulent combustion. "Turbulence modelling" is the euphemism for any set of bald-faced hypotheses that serve to achieve closure, i.e., to reduce the number of unknowns to the number of equations. Theory provides little direction for achieving closure; mainly it establishes criteria, e.g., realizability (nonnegative probability-density functions, etc.) that the results of closure must produce, but it does not show how to satisfy these criteria. Justification for closure hypotheses today remains primarily an experimental topic. Closures usually are traditional; they are based on what has gone before. Fundamentally, experimental results may be consistent with closure hypotheses, but they never can prove them to be correct. There are very few experiments for justification of closures of equations describing the evolution of probability-density functions. By contrast, for some other approaches to be discussed later that achieve closure at a less fundamental level, there is a large body of data consistent with the hypotheses currently employed.

After closure is achieved, a set of equations is obtained, differential equations or integrodifferential equations with independent variables[1] $\mathbf{x}$, t and $\mathbf{v}$, for the evolution of the probability-density functions. The principal activity in the use of this approach then becomes one of obtaining solutions to these equations for particular problems. Usually these problems need initial and boundary conditions for $P(\mathbf{v})$ if they are to be well-posed. It has been indicated that in turbulence statistical properties of the flow tend to be independent of statistics applied at boundaries. To reproduce this behavior, one property of the closure must be that the solution for $P(\mathbf{v})$ is insensitive to certain imposed boundary conditions. The approach has not been pursued far enough to ascertain well whether this property is shared by existing formulations. Most of the problems addressed possess a strong degree of spatial homogeneity, and sometimes stationarity as well, to minimize the need for initial and boundary conditions. Because of the large number of independent variables, computers generally are employed in solving the evolution equations for $P(v)$. In this vein, a considerable amount of research has been devoted to computational algorithms [10]. Results of analyses are available mainly for comparatively simple flows.

Methods based on evolution of probability-density functions seem to be more popular in turbulent combustion than in other areas of turbulence. The reason is that the chemical source term in the equations for species conservation is less troublesome with this type of approach than with other approaches. Severe difficulties that other approaches encounter stem from the highly nonlinear dependence of the source term on $\mathbf{v}$.[2] These difficulties are alleviated greatly when $\mathbf{v}$ is an independent variable. The important thing is that the source term is a one-point quantity; at a fluid level of description rates of chemical reactions depend only on properties at the point in question. The turbulence modelling hypotheses needed in one-point closures for calculating evolutions of probability-density functions then do not involve the source term; rather, they involve transport effects—terms involving spatial gradients. If the approach can be made to work well for nonreactive turbulent-diffusion problems, then it seems likely to work also for turbulent combustion so long as additional closure approximations are not introduced in these latter problems (and they need not be).

As a simple illustration of these conclusions, consider an isothermal, constant-density, constant-diffusivity system in which the velocity field $\mathbf{v}$ is

---

[1] This corresponds to the most usual closures, at the one-point, one-time level, i.e., when all of the components of $\mathbf{v}$ refer to the same position and time. For multipoint closures, for example—say involving L different points—$\mathbf{x}$ here becomes a vector in $3L$ dimensions. Although two-point closures offer some fundamental advantages in potentially reducing questionable aspects of modelling, relatively little research has been pursued on these.

[2] Here $\mathbf{v}$ includes temperature and species concentrations. Below and in all subsequent sections it refers only to velocity.

specified deterministically, and seek an equation for evolution of $P(Y_1, Y_2, \cdots, Y_{N-1})$, the joint probability-density function for the $N-1$ independent mass fractions that satisfy the conservation equation preceding equation (20) of Chapter I. For this P it can be shown that

$$\frac{\partial P}{\partial t} + \mathbf{v} \cdot \nabla P + \sum_{i=1}^{N-1} \frac{\partial}{\partial Y_i} \frac{\dot{\rho}_i}{\rho} P$$

$$= - \sum_{i=1}^{N-1} \frac{\mu_{ii}}{\rho} \frac{\partial}{\partial Y_i} \left[\lim_{\mathbf{x}' \to \mathbf{x}} \nabla^2_{\mathbf{x}'} \int_0^\infty Y_i' P_2(Y_i', Y_1, Y_2, \cdots, Y_{N-1}) \, dY_i' \right]$$

(1) $\quad = \sum_{i=1}^{N-1} \frac{\mu_{ii}}{\rho} \nabla^2 P$

$$- \sum_{i=1}^{N-1} \sum_{j=1}^{N-1} \frac{\mu_{ii}}{\rho} \frac{\partial^2}{\partial Y_i \partial Y_j} \int_{-\infty}^{\infty} X_{ij} P(X_{ij}, Y_1, Y_2, \cdots, Y_{N-1}) \, dX_{ij},$$

where P_2 denotes a two-point joint probability-density function, with the random variable corresponding to Y_i' being that at the point $\mathbf{x}'$, different from $\mathbf{x}$. The subscript $\mathbf{x}'$ on ∇^2 indicates that the coordinates for the operator are $\mathbf{x}'$ rather than $\mathbf{x}$; X_{ii} is a probability-density argument for a random variable that is the dot product of mass-fraction gradients $\nabla Y_i \cdot \nabla Y_j$, and $P(X_{ij}, Y_1, Y_2, \cdots, Y_{N-1})$ is a one-point joint probability-density function for all independent mass fractions and the particular dot product of gradients. The derivation of (1) is too long to be given here but may be inferred from pp. 187–196 of E. E. O'Brien's section of [2]. The result can be understood by analogy with the Liouville equation or the Boltzmann equation of the kinetic theory of gases. The term involving reaction rates describes transport in Y_i space by chemical kinetics and is analogous to transport in velocity space by external forces on molecules. The right-hand side, two forms of which are shown, is loosely analogous to the collision-integral term in the Boltzmann equation. Here this term arises from molecular diffusion and, irrespective of the form adopted, necessitates modelling of a joint probability-density function to achieve closure, basically because diffusion involves gradients, thereby leading to the occurrences of P's different from those on the left-hand side. In contrast, the reaction terms are seen to be retainable exactly, without modelling.

Although an approach based on evolution of probability-density functions thus is extendable readily to include chemistry, it may not be very good for nonreactive turbulence. Its proponents extol the highly fundamental level at which closure is achieved. But good guidelines are lacking for thoroughly evaluating closures at this level. It seems likely that various experimentally known properties of nonuniform turbulent flows will not be reproduced with the closures currently fully available. More research in the area is needed before this speculation can be tested or the approach can be used routinely, with confidence in its predictions.

2.2. Direct numerical integrations for realizations. An approach that in principle does not require any closure approximations is to integrate the full set of conservation equations numerically for an initial-value problem. Unthinkable twenty years ago, with continuing advances in capabilities of computers this approach to analyses of turbulent flows holds increasing promise. Already inviscid (Euler) equations describing buoyant rise of a turbulent plume have been solved without modelling [17], for example, and a random-vortex method has been used to calculate the two-dimensional turbulent flow in a combustor of simple shape [18]. With the advent of supercomputers, and with the potential development of dedicated computers ("hard-wired" to solve the conservation equations), prospects abound for addressing more complicated problems in the future.

Expectations of achievements by numerical approaches of this type must be tempered with recognition of how difficult the tasks really are. Thus, although it is known that three-dimensionality is a vital aspect of the turbulence familiar in combustion (e.g., in profoundly influencing vortex dynamics), with rare exceptions existing programs are strictly two-dimensional. For the Euler equations useful three-dimensional algorithms have been developed. Random-vortex methods always need approximations of some kind at the smaller scales (reminiscent of multigrid concepts), and the problems of extending them to three dimensions are so horrendous that more prosaic finite-difference or finite-element methods (or perhaps spectral methods) may well prove to be more efficient in the long run. Numerical approaches often are thought of as brute-force methods, but in turbulence this hardly seems justified in view of the finesse currently needed to make them work well for problems of interest. Although these approaches have a bright future, the time scale for notable progress seems more like twenty years than five.

A significant problem with numerical-integration approaches concerns what to do with the results once they are obtained. An application of a totally effective method of this type in essence completely reproduces an experiment. Thus, the output possesses a tremendous amount of information—all that is contained in an experiment, in principle. Moreover, unlike the experiment, the information should be readily retrievable. In science, progress in understanding is achieved through processes of abstraction and idealization. The numerical approaches are not designed for this; idealizations, of course, can be put into them, but the investigator must decide what idealizations to make because the method will not. For example, we may look at the measured and calculated rates of growth of a turbulent jet, see that they agree and briefly be happy—but a nagging uneasiness sets in. Is that all there is? Shouldn't other questions be asked? What questions? Theories based on different approaches help by pointing out many interesting and significant questions. Perhaps the development of numerical-integration methods should best be viewed as building a laboratory in which experiments can readily be run to address whatever questions theories may deem desirable.

2.3. Approximations of probability-density functions. An approach that employs probability-density functions but is less ambitious than calculating their evolution is to focus effort on approximating them by use of less fundamental formulations, e.g., the moment methods discussed below. This is an approach that is peculiar to chemically reacting flows and has little to recommend it for nonreacting turbulence. Without chemistry the information provided by the approximated probability-density functions is no greater than that already obtained in approximating them in the first place. But for many reacting flows they provide additional information of considerable interest. Since the utility of the technique relates to specific questions that are asked, further discussion of it is postponed until §4.

2.4. Moment methods. In the development of approaches based on moment methods the starting point is the conservation equations themselves. First they are averaged—in general the average that is taken is an ensemble average. For statistically stationary flows time averages may be employed (and often are, for conceptual simplicity), or, especially if stationarity is absent, appropriate space averages may be introduced if there is any degree of statistical spatial homogeneity in the problem. Since the conservation equations are nonlinear, the differential equations obtained for the averages contain averages of products and (with typical chemistry) of nonlinear functions of the dependent variables. Since the averages of products are unknown, it is conventional to subtract the conservation equation for an averaged quantity from its original conservation equation, multiply by its fluctuation (the difference between it and its average) and average again to obtain a differential equation for the mean square fluctuation, a second moment of the probability-density function. In the equation so obtained for this average of a square (a double product) appear averages of triple products, for example. No matter where the procedure of multiplication and averaging is terminated, there always remain more unknowns than equations. In fact, the whole idea is distinctly unappealing because the farther the procedure is carried, the larger becomes the ratio of the number of unknowns to the number of equations. This is the manifestation of the closure problem as experienced by moment methods, methods in which the investigator works only with moments of probability-density functions.

To illustrate, consider a stationary flow described by equation (2) of Chapter I with ρ constant. Its average is

(2) $$\rho\bar{\mathbf{v}} \cdot \nabla\bar{\mathbf{v}} = \nabla \cdot \bar{\Sigma} - \nabla \cdot (\overline{\rho\mathbf{v}'\mathbf{v}'}),$$

where $\mathbf{v}' = \mathbf{v} - \bar{\mathbf{v}}$. The Reynolds stress, $\overline{\rho\mathbf{v}'\mathbf{v}'}$, is a second moment of $P(\mathbf{v})$. By the indicated manipulations

(3) $$\bar{\mathbf{v}} \cdot \nabla(\overline{\rho\mathbf{v}'\mathbf{v}'}) = \overline{(\mathbf{v}'\nabla \cdot \Sigma)} + \overline{(\mathbf{v}'\nabla \cdot \Sigma)^T}$$
$$- \overline{\rho\mathbf{v}'\mathbf{v}'} \cdot \nabla\bar{\mathbf{v}} - (\overline{\rho\mathbf{v}'\mathbf{v}'} \cdot \nabla\bar{\mathbf{v}})^T - \nabla \cdot (\overline{\rho\mathbf{v}'\mathbf{v}'\mathbf{v}'}),$$

in the last term of which appears the third moment, a tensor with 27 components in contrast to the 9 of the Reynolds stress. The problem is becoming more and more complicated.

Turbulence modelling is introduced to relate higher moments to lower moments so that a closed formulation is obtained in which the number of unknowns equals the number of equations. At the simplest level only the averaged conservation equations themselves are used, and the turbulent transport (e.g., Reynolds stresses) and chemical source terms require modelling. For the former, diffusion approximations (involving guessed formulas for turbulent diffusivities, like that obtained by Prandtl's mixing-length theory) may be employed, while for the latter either averages of rates may be equated to rates evaluated at averages, or approximations motivated by eddy concepts may be introduced (as in so-called "eddy-breakup" models) [2]. Thus, in terms of a turbulent diffusivity ν_T, the Reynolds stresses are

(4) $$\overline{\rho\mathbf{v}'\mathbf{v}'} = -\rho\nu_T[\nabla\bar{\mathbf{v}} + (\nabla\bar{\mathbf{v}})^T],$$

and with a representative form of an eddy-breakup model,

(5) $$\bar{\rho}_i = -\rho C \sqrt{\overline{Y_i'^2}}\, \epsilon/k,$$

in which $Y_i' = Y_i - \bar{Y}_i$, C is an empirical nondimensional constant (a fudge factor), ϵ denotes the average rate of dissipation of turbulent kinetic energy and $k = \overline{(\mathbf{v}' \cdot \mathbf{v}')}/2$ represents the average turbulent kinetic energy. The idea behind (5) is that the reaction rate may be proportional to the rate of dissipation of turbulent kinetic energy in highly turbulent flows. At the next most simple level, differential equations are retained for k and for a quantity, such as a turbulence length scale, related to ϵ. The modelling approximations introduced at this level comprise the popular k-ϵ modelling. Procedures in which differential equations are developed for all second moments, including Reynolds stresses, usually are termed second order closures. Considerations of closures beyond second order are rare. Reviews are available [2], [4], [19], [20] giving the various formulations and closures.

Moment methods have a long and rather successful history of use for constant-density, nonreacting turbulent flows. However, even without chemistry, variable fluid density ρ introduces complications. For example, already in conservation of mass $\nabla \cdot (\rho v)$ appears, which involves a product of two dependent variables unless the mass flux ρv is taken as a dependent variable instead of the velocity v. A way to circumvent the variable-density complication is to work always with mass-weighted averages, which are termed Favre averages since Favre first wrote the moment equations that arise in doing so. With a superscript tilde denoting a Favre average, $\tilde{v} = \overline{(\rho v)}/\overline{\rho}$, and the Favre fluctuation is $v'' = v - \tilde{v}$, obeying $\overline{\rho v''} = 0$, average mass conservation is then simply

$$\text{(6)} \qquad \frac{\partial \overline{\rho}}{\partial t} + \nabla \cdot (\overline{\rho}\tilde{v}) = 0$$

but would involve $\nabla \cdot \overline{(\rho' v')}$ without mass-weighted averaging. Moment-method formulations with Favre averaging always give fewer terms in the equations, except with respect to molecular transport, which involves modelling in these approaches anyway. In a continually growing school the successful constant-density modelling is assumed to be extendable to variable-density flows through Favre averaging. This appears to be a reasonable working hypothesis until more thorough experimental information becomes available on variable-density turbulent flows for making careful checks.

There is a class of problems in turbulent reacting flows for which moment methods are useful and comparatively well justified. These are problems in which fluctuations in temperature and in concentrations of chemical species are sufficiently small in comparison with their local average values. The dependence of the rate of a chemical reaction on temperature T typically is expressed well by an Arrhenius factor $e^{-E/RT}$, where R is the universal gas constant and E the activation energy for the reaction (Chap. I, §2). Unless special modelling is introduced for the chemical source terms, associated with moment methods are expansions of the Arrhenius factor in powers of $T'/\overline{T}$, where $T' = T - \overline{T}$ is the fluctuation about the mean temperature $\overline{T}$. The coefficients in the expansion contain factors $E/R\overline{T}$, which although small for some chemical steps generally are large, 10 to 100, for many of the important ones. The result is that if $T'/\overline{T}$ is of the order unity, then each successive term in this expansion,

$$\text{(7)} \qquad \begin{aligned} \exp(-E/RT) &= \exp(-E/R(\overline{T} + T')) \\ &= \exp\left(-E/R\overline{T} \sum_{n=0}^{\infty} (-1)^n (T'/\overline{T})^n\right) \\ &= \exp(-E/R\overline{T}) \\ &\quad \times \left(\sum_{m=0}^{\infty} \frac{(-1)^m}{m!} (E/R\overline{T})^m \left[\sum_{n=1}^{\infty} (-1)^n (T'/\overline{T})^n\right]^m\right) \end{aligned}$$

is larger than its predecessor by a factor of about ten, until roughly the twentieth term in the expansion is reached. Since twentieth-order moment methods are impractical, $T'/\overline{T}$ must be small, less than a value between 0.01 and 0.1, for the moment approach to be good without special source-term modelling. In turbulent combustion $T'/\overline{T}$ is seldom that small.

Although this trouble often can be remedied by modified modelling of the chemical source term, any such modelling is liable to conceal some of the aspects of interactions between chemical kinetics and the turbulence, and what is worse, a more insidious difficulty lurks beneath. It has recently been established that the chemistry interacts with turbulent transport modelling because, roughly speaking, a fluid element may be transported by turbulence to a location, react there and then be transported back; this process is not taken into account in transport modelling. The phenomenon may cause changes in signs of turbulent diffusivities within the interior of the flow [21]–[24]. Thus, in the popular modelling approximation

$$(8) \qquad \overline{\mathbf{v}'Y_i'} = -D_{Ti}\nabla \overline{Y_i},$$

where D_{Ti} is a turbulent diffusion coefficient, negative values of D_{Ti}, representing "countergradient diffusion," have been found to occur both experimentally [22], [24] and for different theoretical reasons [21], [23] throughout substantial regions of turbulent flames. The now well established fact of countergradient diffusion cannot be handled well by moment methods.

Fortunately, countergradient diffusion does not arise in all turbulent combustion problems. In particular, if there is no chemical source term, then the most fundamental mechanism for it disappears. Atoms are rearranged but neither created nor destroyed by chemical reactions, and the sum of the thermal and chemical enthalpies also is conserved in this manner. Thus, atom (or element) concentrations and total enthalpies are conserved scalars, scalar fields lacking a chemical source term. There is not yet any evidence for countergradient diffusion of conserved scalars. Therefore moment-method formulations may be applied to the conservation equations for conserved scalars. It is then necessary to have different approaches for obtaining solutions for quantities that are not conserved in this sense. These approaches involve the approximation of probability-density functions, mentioned earlier.

2.5. Methods not derived from fluid-level conservation equations. The complexity of turbulent combustion problems has motivated the development of approaches in which the underlying conservation equations for the fluid are abandoned, in a sense. Of course, fluid dynamics is an essential part of the problem and cannot be totally ignored. In fact, it appears in various ways in these approaches, but not at the level of the Navier–Stokes equations. Instead, for example, empirically motivated correlations for mixing rates will be introduced into equations that focus largely on chemical aspects, or the chemistry will be viewed as occurring in coherent structures, flow structures such as vortices that are convected and deformed by the turbulent flow but that maintain their identity over their lifetime. Even for nonreacting turbulence, postulation of conservation equations for coherent structures is a type of approach to which some thought has been given recently. For reacting flows the emphasis has been on the convection, mixing and reaction of the chemical species. There are age theories in which probabilistic methods are applied to describe the age of a fluid element, i.e., the length of time that it has been in a chemical reactor [25]. Still under development [26], [27] is an ESCIMO theory, an acronym for engulfment, stretching, coherence, interdiffusion and moving observer, which identifies parcels of fluid called "folds," having both "demographic" and "biographic" attributes, the former being described probabilistically, somewhat as with age theories, and the latter being fundamentally a deterministic history of chemical transformation within a fold.

An example of an age theory is provided by the problem of finding the probability-density function $P(Y_i)$ for a reactant i at the outlet of a reactor in terms of its probability-density function $P_0(Y_i)$ at the inlet. An equation adopted for this purpose is

$$(9) \qquad \frac{\partial P}{\partial t} = (P_0 - P)/\bar{\tau}_r - \frac{\partial [P\dot{\rho}_i/\rho]}{\partial Y_i} + \psi,$$

where $\bar{\tau}_r$ is a mean residence time in the reactor and ψ, an integral over Y_i quadratic in P, accounts for coalescence and dispersion processes associated with turbulent mixing. A rough similarity is evident between (9) and (1). Modelling is needed for $\bar{\tau}_r$ and for the form of ψ and constants therein. Questions involving $\bar{\tau}_r$ are addressed through experiments involving time-dependent inlet injection of a nonreactive tracer. If $f(\tau_r)$ is a probability-density function for the residence time τ_r, then to interpret the tracer experiment the formula

$$(10) \qquad Y_i(t) = \int_0^t Y_{i0}(t - \tau_r) f(\tau_r) \, d\tau_r$$

may be selected, and an average residence time is

$$(11) \qquad \bar{\tau}_r = \int_0^\infty \tau_r f(\tau_r) \, d\tau_r.$$

If the inlet (Y_{i0}) and outlet (Y_i) concentrations of the nonreactive tracer are measured as functions of time, then in principle $F(\tau_r)$ is readily obtained through inversion of the convolution, although in practice experimental accuracies often are poor. Limiting behaviors are the plug-flow reactor, $f(\tau_r) = \delta(\tau_r - \bar{\tau}_r)$ and the perfectly stirred reactor, $f(\tau_r) = e^{-\tau_r/\bar{\tau}_r}/\bar{\tau}_r$.

While approaches in this category have attractive aspects, such as the ability to address certain questions that are too difficult for other theories, their prospects for the future elude assessment. Although they certainly overlook many phenomena, their level of entry makes it difficult to see just what has been left out.

2.6. Perturbation methods. In some turbulent flows it is possible to avoid modelling hypotheses entirely through the use of formal expansion procedures for limiting values of certain parameters that characterize the flow. Such expansions sometimes are perturbations of laminar flows and therefore may apply only for relatively limited ranges of turbulent conditions. Examples are expansions for low turbulence intensities or for large turbulence scales and expansions about local chemical equilibrium [2]; specific examples will be encountered later. Asymptotic expansions recently have been introduced with some success to take advantage of disparities of scales that often occur in turbulent combustion [11]. Within the ranges of validities of the expansions, the predictions possess high degrees of confidence—higher than those achievable even in principle by any other types of techniques except direct numerical integrations. Therefore under appropriate circumstances results obtained by perturbation methods may provide standards by which accuracies of other types of techniques may be calibrated.

A good underlying philosophy for developing theories of turbulent combustion is to attempt to reduce the problem addressed to one involving a nonreactive turbulent flow. Turbulence is tough enough, even without chemistry, and putting combustion into it certainly does not help to simplify it. If the combustion can be gotten out of the turbulent combustion problem in a well-justified manner, e.g., by working with variables like conserved scalars and by introducing results obtained by perturbation methods, then the best available methods for nonreacting turbulent flows can be applied to complete the solution of the turbulent combustion problem. Many of the studies of turbulent combustion by perturbation methods subscribe to this objective.

3. Regimes. Another good objective in initiating a study of turbulent combustion is to first identify as well as possible the regime that is to be considered. In different regimes turbulent reacting flows exhibit very different behaviors. Identification of regimes is based on values of relevant nondeimensional parameters. Therefore parameters of turbulent combustion must first be introduced.

3.1. Relevant parameters. Parameters of both chemistry and turbulent flow are involved in identifying regimes. Since there are numerous such parameters [28], especially concerning the chemistry, considerable simplification is needed to obtain a comprehensible classification. Here the chemistry will be characterized by just one parameter, a representative reaction time τ_c. Since many reaction steps occur in turbulent combustion, τ_c pertains not to one step but rather to the overall effect of many steps in releasing heat. Even with this simplification, τ_c varies strongly with local conditions, e.g., with T, and therefore to obtain a relevant constant, τ_c will be evaluated in the hottest part of the combustion field. The ratio of representative fluid time to τ_c is a Damköhler number (see Chapter I), a dimensionless parameter first introduced by Damköhler [29] for describing reacting flows. There are many Damköhler numbers, depending on the selection of the fluid time. A large-eddy Damköhler number D_l may be defined by selecting a representative turnover time of a large eddy, $l/\sqrt{2k}$, where l is the so-called integral scale of the turbulence. Thus $D_l \equiv (l/\sqrt{2k})/\tau_c$ measures the extent to which the chemistry can proceed to completion in a large eddy.

A parameter relevant to the fluid dynamics of the turbulence is the turbulence Reynolds number, $R_l \equiv \sqrt{2k}\, l/\nu$, where ν is the kinematic viscosity. The molecular transport coefficient ν also varies appreciably with T, and therefore to obtain a constant value here, it too will be evaluated in the hot gas. Since gases are of interest here, for order-of-magnitude estimates, all molecular diffusivities, i.e., those for momentum, energy and chemical species, will be taken equal. With this selection, the turbulence properties for all of the fields can be characterized solely in terms of R_l. Thus, the regimes of turbulent combustion can be represented in a plane with coordinates R_l and D_l.

Although R_l and D_l suffice to characterize the regimes of turbulent combustion under the approximations that have been introduced, different parameters often are employed to emphasize various physical aspects. These other parameters always can be related to R_l and D_l. Prior to definition of some of the other parameters it will be necessary to distinguish two types of turbulent combustion on the basis of the initial arrangement of reactants.

3.2. Premixed and nonpremixed systems. A first instinct in visualizing reacting flows is to consider a reactant that can be converted into a product. The extent of conversion may be thought of as measured by a reaction progress variable Y, allowed to vary from zero for pure reactant to unity for pure product. Here Y may be viewed as the mass fraction of product in the mixture, although there are particular problems in which mass-flux fractions afford better definitions. With the former selection, a conservation equation for Y roughly of the form

(12) $$\frac{\partial}{\partial t}(\rho Y) + \nabla \cdot (\rho \mathbf{v} Y) = \nabla \cdot (\rho \nu \nabla Y) + w$$

may be derived, where w denotes an appropriately normalized rate of the chemical reaction (compare Chapter I, §1). In a turbulent flow, where Y becomes a stochastic function, a probability-density function for Y, at any fixed x and t, is usually denoted simply by $P(Y)$.

In combustion the chemistry seldom involves only one reactant. Typically there are two, fuel and oxidizer, which may or may not be mixed before a combustion experiment is initiated. If there is thorough mixing in advance, then the reacting flow is said to be a premixed system, or a premixed flame, and for many purposes may be treated as if there were only one reactant; a single variable Y may describe the chemical composition. Often the fuel and oxidizer are supplied in separate streams and mix in the flow as the chemistry proceeds—these flows are said to be nonpremixed systems or diffusion flames since diffusion then manifestly is essential to combustion. For nonpremixed systems an additional variable is needed to describe the chemical composition, viz., a measure of the relative amounts of fuel-containing and oxidizer-containing species present. A convenient choice for this variable is the mixture fraction Z, the ratio of fuel mass (irrespective of the molecules, reactants or products, in which the fuel atoms may be contained) to the total mass in any small volume element; thus Z is defined to go from zero in the oxidizer stream to unity in the fuel stream. Since Z is a conserved scalar, it obeys a conservation equation simpler than that for Y, viz.,

$$(13) \qquad \frac{\partial}{\partial t}(\rho Z) + \nabla \cdot (\rho \mathbf{v} Z) = \nabla \cdot (\rho \nu \nabla Z).$$

The probability-density function $P(Z)$ is useful in nonpremixed turbulent combustion. For each value of Z the progress variable Y may be defined to vary from zero, if the fuel and oxidizer have mixed without reacting at all, to unity, if the reaction has proceeded all the way to chemical equilibrium at the given value of Z. Thus, in the nonpremixed system the composition may be represented as a point in the plane of the two variables, Y and Z, each in the range zero to unity. The joint probability-density function $P(Y,Z)$ (at the same $\mathbf{x},t$) may be of interest in turbulence.

These definitions suggest that turbulent combustion in nonpremixed systems is more complicated than that for premixed systems. Surprisingly, this is untrue in many practical respects. Because of variations in Z, the limit of complete chemical equilibrium, $Y = 1$, is nontrivial for nonpremixed systems. In this limit $[P(Y, Z) = \delta(Y - 1)P(Z), P(Z|Y = 1) = P(Z)]$, the chemical composition depends only on the conserved scalar Z, so that the previously cited difficulties associated with the chemical source term are avoided. The turbulent combustion problem in fact becomes equivalent to a problem in nonreactive turbulence for mixing of two fluids with somewhat complicated caloric equations of state (equations relating enthalpy to temperature and pressure). For premixed systems, chemical equilibrium ($Y =$

1) would effectively give a turbulent flow problem for a pure, nonreactive species, a problem that would have nothing to do with turbulent combustion. Nonequilibrium somewhere is an essential aspect of combustion only for premixed systems.

The mixture fraction Z is a sufficient variable only in a limited class of nonpremixed systems. There may be three-reactant systems, three-stream problems for two-reactant systems with different chemical compositions in each inlet stream, partially premixed two-reactant, two-stream problems, etc., each of which needs variables beyond just one Z for full specification of composition. However, many of the interesting nonpremixed systems have only two types of streams, identifiable as fuel and oxidizer. The distinction between premixed and nonpremixed combustion provides an important classification, in some respects the most basic of all, for both laminar and turbulent flows. Thus, characteristics of the regimes of turbulent combustion can best be discussed separately for diffusion flames and premixed flames.

3.3. Additional parameters. From ν and τ_c a velocity associated with chemical kinetics and molecular diffusion may be constructed as $v_0 = \sqrt{\nu/\tau_c}$. Although this quantity exists for both premixed and nonpremixed systems, it has an important physical meaning only for premixed flames. It is the velocity at which a laminar flame propagates into a combustible mixture, the laminar flame speed (or burning velocity) [30], as discussed in Chapter I, §4. It provides a velocity, related to chemistry, against which turbulence velocities can be compared. Thus, $\sqrt{2k}/v_0$ may be taken to represent a nondimensional turbulence intensity, one that may be employed for all turbulent combustion problems but possesses the most direct physical significance for premixed systems. Observe from the definitions that $\sqrt{2k}/v_0 = \sqrt{R_l/D_l}$.

Associated with v_0 is a length, $\delta = \nu/v_0$, the thickness of a premixed laminar flame. Turbulence length scales may be compared with δ; e.g., $l/\delta = \sqrt{R_l D_l}$ is a nondimensional turbulence scale. For $R_l > 1$ there is a distribution of turbulence length scales [31] ranging from the largest, of order l, to the smallest, of the order of the Kolmogorov scale, $l_k = (\nu^3/\epsilon)^{1/4} = l/R_l^{3/4}$. If there is interest in comparing the smallest turbulence scales with the thickness of a laminar flame, then $l_k/\delta = \sqrt{D_l}/R_l^{1/4}$ is a relevant parameter. Although the rate of dissipation of turbulent kinetic energy is greatest at the Kolmogorov scale, there is an intermediate scale associated with the average rate of dissipation, the Taylor scale l_t, appearing in $\epsilon = (2k)^{3/2}/l = \nu(2k)/l_t^2$, and therefore given by $l_t = l/\sqrt{R_l}$. Reynolds numbers may be associated with scales other than l; for example, the Reynolds number based on the Taylor scale is $R_t \equiv \sqrt{2k}\,l_t/\nu = \sqrt{R_l}$. Damköhler numbers may be associated with turbulence times other than the large-eddy turnover time; for example, a characteristic time for a Kolmogorov eddy is $\tau_k = \sqrt{\nu/\epsilon}$, and the corresponding Damköhler number, $D_k = \tau_k/\tau_c$, is readily shown to equal $(l_k/\delta)^2$. The relationships among various nondimensional parameters are plotted in Fig. 2, which is similar to a graph given earlier [11] and basically equivalent to one first prepared by K. N. C. Bray for his contribution to [2].

3.4. Identification of regimes. In Fig. 2 the graph with coordinates l_k/δ and $\sqrt{2k}/v_0$ is more appropriate for premixed flames while that with coordinates R_l and D_l seems better for diffusion flames, although the two fundamentally are equivalent and show the same information.

In the region of the first graph marked "decay of turbulence," which has been excluded from the second, the decay time for turbulence is less than the transit time through a laminar flame. This region is not very interesting because the turbulence exerts relatively little influence there. The most important limiting regimes are those marked "stirred reactors" and "reaction sheets." The first applies to high-intensity, small-scale turbulence and the second to large-scale turbulence of any intensity. The first tends to occur at small Damköhler numbers and the second at large Damköhler numbers. Whether precise locations of boundaries between these two limiting regimes exist is not known today, but the two limits are known to exist and to exhibit quite different characteristics [11].

In the stirred-reactor regime the turbulent mixing is fast in comparison with the chemistry, and reaction therefore tends to occur in broadly distributed reaction zones, as encountered in well-stirred chemical reactors, for example. In the reaction-sheet regime the chemistry in hot fluid elements is fast in comparison with turbulence processes. Since the reaction rates depend strongly on temperature, the chemistry need not be fast in cold fluid elements; in premixed flames fundamentally it cannot be, and in diffusion flames typically it also is not. Therefore in the reaction-sheet regime the chemistry occurs only in the hot portions of the flow, which tend to form thin sheets, convected and distorted by the turbulence.

The extent of sheet distortion in the reaction-sheet regime increases with increasing turbulence intensity. With increasing distortion the sheets fold over upon themselves and eventually may become highly convoluted. For premixed flames there is a simple mechanism (laminar flame propagation) by which different portions of sheets approach each other and cut off pockets of unburnt gas, thereby forming multiply connected reaction surfaces. A tentative boundary between multiple sheets and single sheets for premixed flames in the reaction-sheet regime is illustrated in the figures. The location of this boundary in fact is uncertain, as are all boundary locations in the

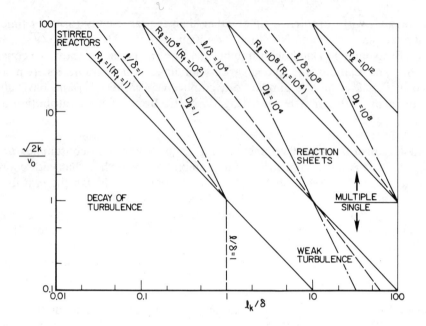
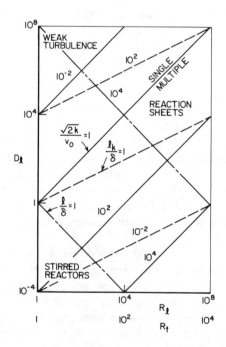

Fig. 2. *Regimes of turbulent combustion.*

plane. There may well exist currently unidentified intermediate regimes, e.g., between $l/\delta = 1$ and $D_l = 1$, or between $D_l = 1$ and $l_k/\delta = 1$, with distinguishing characteristics of their own.

Development of well-justified analyses of turbulent combustion is likely to be easiest in the weak-turbulence limit of the reaction-sheet regime, indicated in the figures. In fact, some progress in this has been made for premixed flames, as indicated below. Approximations of low turbulence intensity need not be accompanied by reaction-sheet approximations. There may exist distinguished types of weak-turbulence behavior that do not possess a reaction-sheet character (e.g., at the lower values of l_k/δ), but this is not yet known.

From this discussion it is seen that graphs suitable for the placement of different regimes of turbulent combustion are available, and certain limiting regimes are known to exist. But there are many unknowns concerning other possible regimes and boundaries between regimes, including possibilities of discontinuities in characteristics across lines in the graphs. In view of these uncertainties, when approaches to analyses of turbulent combustion are pursued, it seems desirable to attempt to identify the regions in the graphs where the approaches are likely to be most reasonable.

4. Assessments. Since two principal limiting regimes of turbulent combustion have been identified, and since combustion processes are known to differ for premixed and nonpremixed systems, in assessing what is known about the subject and useful approaches to the analysis thereof, it seems best to discuss separately each of the four combinations.

4.1. Diffusion flames with reaction sheets. For diffusion flames the reaction-sheet limit corresponds most simply to conditions of full chemical equilibrium which, as has been stated, can be approached on the basis of a nonreacting flow having complicated thermodynamics. Under these conditions the main advantage of an approach based on evolution of probability-density functions (i.e., the ability to extend the method readily to handle chemical kinetics) goes away; the principal difficulty with that approach pertains to the mixing problem, and the problem at hand is one of mixing. Considerable success has been achieved here through approximation of the probability-density function $P(Z)$ (see R. W. Bilger's contribution to [2]). The shape of $P(Z)$ is parameterized with two to four parameters, typically as a smooth curve between $Z = 0$ and $Z = 1$, with delta functions at one or both of its ends to account for intermittency (here meaning the presence of pure fuel or pure oxidizer in some of the fluid elements in the ensemble). The parameters are expressed in terms of moments ($\overline{Z}$, the mean square fluctuation of Z, etc.), and from moment methods applied to the conserved scalar Z the parameters may then be calculated as functions of position and/

or time. Flow properties of interest, such as the temperature T, are directly related to Z at equilibrium, so that $P(T)$, for example, may be obtained from $P(Z)$, and desired properties of the T field, e.g., $\overline{T}$, may be calculated. This approach provides a considerable improvement over use of moment methods directly for T because of modelling difficulties, e.g., countergradient diffusion, in the moments of the T field, which possesses a chemical source term. It is the nonlinearity of the equilibrium relationship between T and Z that necessitates introduction of $P(Z)$ in this approach. There are a number of finer points involved, which have been covered in earlier reviews [2] [11].

To illustrate the approach, consider the beta-function distribution,

(14) $$P(Z) = Z^{\alpha-1}(1 - Z)^{\beta-1}\Gamma(\alpha + \beta)/[\Gamma(\alpha)\Gamma(\beta)],$$

for $0 \leq Z \leq 1$, with $P(Z) = 0$ otherwise. The nonnegative constants α and β are then related to $\overline{Z}$ and $\overline{Z'^2}$ by

(15) $$\overline{Z} = \alpha/(\alpha + \beta), \qquad \overline{Z'^2} = \overline{Z}(1 - \overline{Z})/(\alpha + \beta + 1).$$

With ρ and ν taken constant for simplicity, the average of (13) for stationary flows is

(16) $$\nabla \cdot (\overline{\mathbf{v}}\overline{Z}) = \nabla \cdot (\nu\nabla\overline{Z} - \overline{\mathbf{v}'Z'}),$$

and a straightforward calculation also gives

(17) $$\overline{\mathbf{v}} \cdot \nabla\overline{Z'^2} = -2\overline{\mathbf{v}'Z'} \cdot \nabla\overline{Z} - \nabla \cdot (\overline{\mathbf{v}'Z'^2}) - 2\nu\overline{|\nabla Z'|^2} + \nu\nabla^2\overline{Z'^2}.$$

In (16) and (17) put

(18) $$\overline{\mathbf{v}'Z'} = -D_T\nabla\overline{Z}, \qquad \overline{\mathbf{v}'Z'^2} = -D_T\nabla\overline{Z'^2},$$

which may be compared with (8), and employ

(19) $$\overline{X} \equiv 2\nu\overline{|\nabla Z'|^2} = \nu\overline{Z'^2}/l_Z^2,$$

where l_Z is a dissipation length for the Z field. Equations (2) and (4) may then be used, along with the additional relationships

(20) $$D_T = \nu_T = k^2/\epsilon, \qquad \epsilon = \nu k/l_Z^2,$$

into each of which an empirical constant is inserted to improve agreement with experiment. Finally, the differential equations of k-ϵ modelling, mentioned earlier, are included to complete the set and enable $P(Z)$ to be calculated at each point through numerical integration. Obtaining $P(T)$, $P(Y_O)$ and $P(Y_F)$ for temperature, oxidizer and fuel then becomes a standard exercise in probability at chemical equilibrium. In the reaction-sheet limit of chemical equilibrium with constant specific heat, the dependence of T, Y_O and Y_F on Z are as illustrated by the solid lines in Fig. 3.

Some relevant questions can be addressed directly by the method that has just been outlined, while others cannot. An important problem concerns the

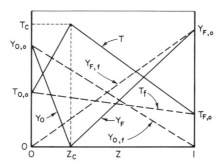

FIG. 3. *Equilibrium and frozen dependences of temperature and fuel and oxidizer mass fractions on the mixture fraction in the reaction-sheet limit.*

production of pollutants, e.g., oxides of nitrogen, in turbulent diffusion flames. Usually this occurs in side reactions that are not germane to the heat release and that are far from chemical equilibrium, and the pollutants, being trace species, affect the combustion negligibly. To the extent that the rates of the side reactions are expressible in terms of concentrations of major species and temperature (as they often are), the average production rates of the pollutants can be calculated from the available probability-density functions, basically from $P(Z)$ [2], [11], [32], [33].

Another property of practical interest is flame length. Suppose that a turbulent fuel jet issues into an oxidizing atmosphere. The distance from the jet exit to the average position at which visible radiant energy emission ceases is the visible flame length. In a useful classical approximation to equilibrium diffusion-flame structure [30], the heat release is localized at the stoichiometric value of Z, say $Z = Z_c$, the mixture fraction at which fuel and oxidizer are present in stoichiometric proportions (the proportions needed for complete combustion with no fuel or oxidizer left over). In this reaction-sheet or flame-sheet approximation, the flame may be considered to be at $Z = Z_c$ (see Fig. 3), so that the instantaneous flame length is the distance in the axial direction from the exit plane to a plane parallel thereto and containing the farthest point at which $Z = Z_c$. Definition of an average flame length then involves an average of this axial distance, which fundamentally is not a one-point quantity and therefore in principle cannot be obtained from $P(Z)$. A zero-crossing problem [16] is involved here, which motivates more detailed statistical investigations that have not been carried out [11].

In the flame-sheet approximation the average rate of heat release per unit volume is another quantity of practical interest that cannot be obtained from $P(Z)$. This has been demonstrated to be related to the joint probability-density function $P(Z, |\nabla Z|)$ (see R. W. Bilger's work [2]). Theoretical and experimental efforts directed toward finding this joint probability-density function have now been in progress for some time and are beginning to bear fruit [2], [11].

The properties identified thus far relate to chemical equilibrium in diffusion flames. Departures from equilibrium occur when $D_l < \infty$. Their consideration fundamentally necessitates the investigation of $Y < 1$, i.e., $P(Y, Z) \neq \delta(Y - 1)P(Z)$. If abrupt transitions do not occur as D_l is decreased, then a continuous merging of regimes of large D_l and small D_l (the stirred-reactor regime, to be considered next) may be expected. Bilger [2] has addressed the algebraically complicated problem of developing perturbation expansions in D_l^{-1}, applicable for Y near unity, for particular chemical-kinetic schemes. Knowledge of when abrupt transitions should be anticipated may be gleaned from analyses of laminar diffusion-flame structure by activation-energy asymptotics [34] (see Chapter I). If chemical equilibrium tends to be maintained to the same extent for all $Z \geqslant Z_c$ (or for all $Z \leqslant Z_c$), e.g., if one of the boundary temperatures exceeds the adiabatic flame temperature T_c (the equilibrium temperature at $Z = Z_c$), then a gradual transition may be expected. In Fig. 3 this corresponds to either $T_{F,0}$ or $T_{O,0}$ being greater than T_c. In most combustion problems this condition is not attained, and abrupt transitions that physically represent flame extinction occur when a laminar Damköhler number falls below a critical extinction value. A review of flame extinction is available [35]. Extension of theoretical results on extinction to turbulent diffusion flames has been made by N. Peters (see [11] for a review).

If there is an abrupt extinction transition, then there should be a line in Fig. 2 across which this transition occurs. No such line has yet been identified. Its location, if it exists, may depend upon the particular turbulent combustion problem considered, e.g., on boundary conditions, and the possibility arises of more than one turbulent-combustion solution at a given point on the graph, e.g., an ignition line, different from an extinction line, with the possibility of both vigorously and weakly reacting flows in between.

In analyses accounting for extinction it is found that the departure of Y from unity is small where extinction occurs [36]. This makes it unnecessary to consider the complications of turbulence modelling in conservation equations for quantities like Y that have chemical source terms. There are very nearly only two flow conditions, reacting ($Y = 1$) and extinguished ($Y = 0$). Local extinction is expressible in terms of $|\nabla Z|$ [36], and therefore a fraction extinguished can be related to $P(Z, |\nabla Z|)$, the same quantity that is relevant to the average rate of heat release. In a sense, there is a maximum rate of heat release per unit volume that a turbulent diffusion flame can support.

These results have a bearing on liftoff and blowoff of turbulent-jet diffusion flames of the type defined above [37]. If the velocity of the fuel jet is increased above a critical value, then experimentally the flame is seen to become detached from the jet exit (liftoff) and to be stabilized in a lifted position away from the jet exit. The distance from the jet exit to the beginning of the flame is the liftoff height, which increases as the jet velocity is increased further. At a higher jet velocity, the critical value for blowoff, the flame disappears entirely and can no longer be stabilized in the jet. Correlations of measured liftoff heights have been obtained from the viewpoint that has been discussed here [37], [38].

The method involves introducing Z of (13) as an independent variable. For a one-step reaction with the chemistry occurring in a narrow range of Z about Z_c, equation (35) of Chapter I when so transformed and stretched in the first approximation becomes simply

$$\frac{\partial^2 T}{\partial Z^2} = -Qw/(\lambda |\nabla Z|^2) \tag{21}$$

for Le = 1, where Qw is the energy per unit volume per unit time released by the chemical reaction. Results of activation-energy asymptotics [34] for laminar diffusion flames then readily provide the maximum value of $|\nabla Z|$ above which extinction of the reaction sheet occurs. At high turbulence Reynolds numbers there is little difference between $|\nabla Z|$ and $|\nabla Z'|$. Therefore, there is a critical value of the instantaneous rate of scalar dissipation for extinction, and liftoff may be characterized approximately by the average $\overline{X}$ of (19). Theories of nonreacting turbulent jets are needed for calculating $\overline{X}$ fields. These fields may be calculated, for example, with analyses like those of (15) through (20). Theoretical results for three different turbulence models are shown in Fig. 4, where measurements from liftoff heights of turbulent-jet methane diffusion flames in air also are shown, with h the liftoff height and d the tube diameter. Reasonable agreements are seen to be obtainable.

4.2. Diffusion flames with distributed reactions. In the stirred-reactor regime of small D_l approaches based on approximating $P(Z)$ no longer seem helpful because, since $Y \neq 1$, the flow properties of interest are not directly related to Z. Here turbulent mixing results in a significant amount of molecular-scale mixing prior to chemical reaction, and finite chemical reaction rates (nonequilibrium) are important in the turbulent combustion process. If there is sufficient spatial homogeneity, then methods based on evolution of probability-density functions can be useful. Moment methods also may be employed, if temperature fluctuation levels are not too large, as indicated earlier. At sufficiently small values of D_l it would appear that for some problems (e.g., reactor flows, but not simple mixing layers) mixing may proceed nearly to completion at a molecular level prior to the occurrence of an appreciable amount of chemical reaction. In this limit the nonpremixed system becomes essentially premixed and shares the properties discussed in the following section.

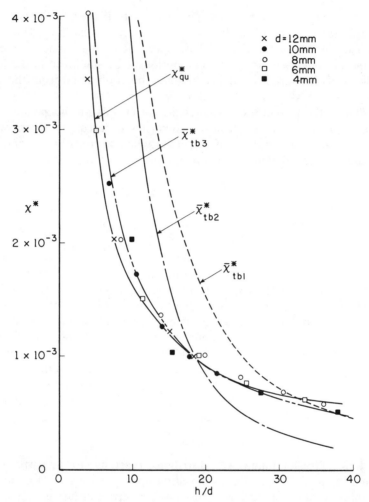

FIG. 4. *Nondimensional rate of scalar dissipation, $\overline{X}^* = \overline{X}d/U$ (U = exit velocity) as a function of the ratio of the liftoff height to the exit diameter for turbulent-jet diffusion flames.*

4.3. Premixed flames with distributed reactions. In premixed systems as well, at low values of D_l the methods identified in the preceding paragraph seem most appropriate. The age theories, mentioned earlier, also address this situation. Recall that since Z is fixed in premixed systems, methods involving approximation of probability-density functions would focus on $P(Y)$. Since there is a source term in the equation for Y, these methods, which have been pursued extensively [2], encounter difficulties. Instead of reasoning on the basis of $P(Y)$, it has been more fruitful in applications to work with averaged conservation equations and equations for low moments by approximating the chemical source term in a rough manner (e.g., on the basis of eddy-breakup ideas, most applicable at high turbulence intensities) [2]. None of the approaches are satisfactory in a fundamental sense, even though they sometimes achieve correlation with experiment. In many respects the high-intensity, small-scale regime is the most difficult one to address in a satisfying way.

An outstanding question in this regime concerns the extent to which a turbulent flame speed (or burning velocity) exists in premixed turbulent flows. If it exists, then the turbulent flame speed v_T, the turbulent analogue of v_0, may be expected to depend on both the chemical nature of the combustible mixture and its turbulence properties. Reasoning that favors the existence of v_T may be given for large-scale turbulence, but the situation for small-scale, high-intensity turbulence is unclear. Within well-stirred chemical reactors there does not seem to be experimental evidence favoring propagating turbulent flames. In confined, high-intensity, premixed turbulent flows flames are observed to spread at an approximately constant angle from bluff-body flame holders, but turbulence modifications by the heat release in these particular, oblique configurations may be responsible for the result; the process is strongly dominated by fluid mechanics, and eddy-breakup models are successful. For open configurations there are no good experiments to provide direction for forming ideas about the high-intensity, small-scale regime. The few concepts that have been put forth in the literature are conjectural.

The possibility of abrupt transitions in Fig. 2 applies equally to premixed systems. The known flame structure at large turbulence scales, discussed in the following section, cannot apply in the stirred-reactor regime. Recognition of this fact has led to suggestions that in open configurations extinction occurs prior to reaching the stirred-reactor regime. Nevertheless, it would seem that, given enough time, the combustible mixture would find some way to react, although perhaps not in a front propagating at a well-defined speed v_T. The current status of understanding of turbulent combustion in the stirred reactor regime seems mainly to be one of confusion.

4.4. Premixed flames with reaction sheets. In contrast, significant advances have been made recently in understanding premixed turbulent-flame propagation in the reaction-sheet regime. Since there are many aspects to the subject, reference must be made to recent reviews [8], [9], [11] for full information.

In the reaction-sheet regime turbulent flames are composed of wrinkled laminar flames. If the turbulence scales are large enough (e.g., $l_k/\delta \gg 1$), then the laminar-flame curvature (associated with the wrinkling) and the strain rate introduced by the turbulence affect the laminar-flame structure to a negligible extent. At smaller turbulence scales there are local modifications to the laminar-flame structure. Although much knowledge concerning these modifications has been accumulated in recent years [9], for clarity it seems best to begin by considering the very large-scale limit in which the modifications are unimportant. In this limit, each element of the wrinkled laminar flame propagates normal to the local flame sheet at velocity v_0 with respect to the cold combustible just ahead of it. This fact enables the turbulent flame speed to be related to the total wrinkled-flame area A_f per unit cross-sectional area of the flow A. As observed by Damköhler in 1940, since all of the fluid flowing through the turbulent flame also passes through the wrinkled laminar flame, $m_T = m_0(A_f/A)$, where $m_T = \rho_0 v_T$ and $m_0 = \rho_0 v_0$ are mass fluxes, or mass burning velocities, from which burning velocities readily are recovered through division by the unburnt gas density ρ_0 (compare (69) of Chapter I). If $G(\mathbf{x}, t)$ is a function that locates the reaction sheet by $G = 0$ with $G > 0$ in the burnt gas and $\mathbf{e}$ is a constant unit vector in the direction of turbulent-flame propagation, then an equivalent expression is

$$(22) \qquad m_T = m_0(\overline{|\nabla G|/|\mathbf{e} \cdot \nabla G|})_{G=0},$$

where the bar is an ensemble average that includes all of the wrinkled-flame positions.

Use of this formula for the turbulent mass burning velocity m_T necessitates evaluating the average, which involves investigating the dynamics of a wrinkled flame in a turbulent flow. With negligible laminar flame-speed modifications the wrinkled-flame motion may be described by the equation

$$(23) \qquad \rho \frac{\partial G}{\partial t} + \rho \mathbf{v} \cdot \nabla G = m_0 |\nabla G|,$$

in which the introduction of m_0 has enabled the equation to be applied in both the unburnt and burnt gas. Since there is a density change across the wrinkled flame, its presence modifies the turbulent velocity fields on both sides of it. These hydrodynamic influences (see Chapter I, §8) are a major source of complication in studies of premixed turbulent flame propagation. G. Darrieus in 1938 and L. Landau in 1944 independently discovered hydrodynamic instability of planar premixed laminar flames; because of the density decrease experienced by a fluid element in passing through the flame, a simple mechanism [9] causes the planar flame to be unconditionally unstable if m_0 is constant. Analyses of motions of wrinkled flames in turbulent flows must cope with hydrodynamic instability.

Instabilities. Instabilities have plagued theoretical studies of premixed turbulent combustion in reaction-sheet regimes. By contrast, since there are fewer instabilities of reaction sheets for diffusion flames, studies of turbulent diffusion flames are relatively free from instability problems. For premixed flames, in addition to the hydrodynamic instability there are buoyant instabilities and diffusive-thermal instabilities, the latter being associated with internal flame-structure modifications involving molecular diffusion of chemical species and conduction of heat (see Chapter I, §10). Unlike the hydrodynamic instability, which always is destabilizing, the buoyant and diffusive-thermal effects may be either stabilizing or destabilizing, depending on the direction of flame propagation and the reactant mixture. It is fair to say that we do not yet have any thorough, unquestionable analysis of premixed turbulent flame propagation for conditions under which the planar laminar flame is unstable. Only if both buoyant and diffusive-thermal effects are stabilizing can they conspire to stabilize the hydrodynamic instability completely (the former handling the large wavelengths and the latter the small). In fact, it is only under this condition that a complete theory exists, and even then only for weak turbulence [9].

A thorough discussion of flame instabilities is beyond the scope of this article (see [30], for example, for more information). At scales smaller than those of diffusive-thermal instabilities are potential chemical-kinetic instabilities within flames, and at scales typically larger than those of buoyant instabilities are acoustic instabilities of flames in chambers. The larger-scale instabilities may interact with turbulence to modify the turbulent combustion. There are experimental conditions under which turbulent-flame accelerations are clearly attributable to instabilities (typically buoyant or acoustic). The hydrodynamic instability, originally thought to be associated with transition to turbulence in flames, seems on the basis of much recent work [8], [9] to instead evolve to rather regular patterns of limited amplitude; it does not appear to proceed to chaos. In this respect, it differs greatly from the fluid-dynamic instabilities that lead to turbulence. Inclusion of effects of hydrodynamic instability in theories of premixed turbulent-flame propagation is a topic of continuing research concerning which further progress may be expected. Up until now most of the studies of premixed turbulent combustion have bypassed hydrodynamic instability, either by considering high turbulence intensities and assuming that they overpower it [39], or by addressing only conditions near the wrinkled laminar flame, leaving studies of far-field regions of hydrodynamic adjustment for future investigations [40].

Flame stretch. Next, let us consider what happens as l_k/δ is decreased somewhat. The internal laminar-flame structure then begins to be affected by the turbulence (of course it might in fact be affected by instability even for $l_k/\delta \to \infty$). In large-scale turbulence, for flames that are at most modified weakly by their instability, the influence of strain and curvature on the laminar-flame structure arises through a single parameter related to a quantity that has been called the "flame stretch." The flame stretch, introduced originally by Karlovitz [41] and later given more general definitions [42], [43], is the time derivative of the logarithm of an area element of the flame sheet, the boundary of the area being considered to move with the local tangential component of fluid velocity at the sheet. For a given flame sheet in a flow there will be a distribution of stretch along the sheet, typically with both positive and negative stretch but with positive values predominating [42]. The relevant nondimensional parameter measuring the extent of modification of the flame structure is a nondimensional version of the total stretch of the flame sheet produced by flow with respect to the moving, curved flame, an effective curvature of the flame.

If $\mathbf{n}$ is a unit vector normal to the reaction sheet, directed from the burnt gas toward the fresh mixture, and $\Phi = [(\nabla \mathbf{v}) + (\nabla \mathbf{v})^T]/2$ is the local rate-of-strain tensor, say measured in the unburnt gas ahead of the flame, then the nondimensional flame stretch is

$$(24) \qquad \kappa = (\delta/v_0)(v_0 \nabla \cdot \mathbf{n} - \mathbf{n} \cdot \Phi \cdot \mathbf{n}).$$

Here the curvature term $v_0 \nabla \cdot \mathbf{n}$ is the stretch that the curved sheet would experience if it were moving in a quiescent medium at the normal burning velocity v_0. To this is added the stretch that a planar flame would experience in the flow, $-\mathbf{n} \cdot \Phi \cdot \mathbf{n} = -\mathbf{n} \cdot \nabla \times (\mathbf{v} \times n) = -(\mathbf{n} \cdot \nabla)(\mathbf{v} \cdot \mathbf{n}) + \nabla \cdot \mathbf{v}$ for a planar flame. The nondimensionalizing factor, δ/v_0, is the residence time of a fluid element in the laminar flame. Since v_0 is inversely proportional to the chemical time τ_c, κ is the reciprocal of a Damköhler number for the stretched laminar flame. If x is not parallel to the flame sheet and the se- lection $G(\mathbf{x}, t) = x - F(y, z, t)$ is introduced, then it can be shown that

$$(25) \qquad \kappa = (\delta/v_0)\left\{\nabla \cdot \mathbf{v}_t + \left[v_0 \nabla^2 F + \left(\frac{\partial}{\partial t} + \mathbf{v}_t \cdot \nabla\right)\sqrt{1 + |\nabla F|^2}\right]\Big/\sqrt{1 + |\nabla F|^2}\right\},$$

where $\mathbf{v}_t$ is the two-dimensional vector representing the transverse (y, z) components of fluid velocity upstream. Geometrical and kinematic aspects of flame stretch have recently been clarified [44]–[47]; there are many other expressions that are equivalent to (24) and (25).

Analyses for weak stretch. To investigate influences of strain rate and reaction-sheet curvature on wrinkled-flame structure and dynamics, it is convenient to begin with conservation equations written in a coordinate system moving with the reaction sheet [9]. Many results recently have been obtained from investigations of this type [8], [9], [11], [21], [40], [45]–[53]. Through perturbation methods the results provide a correction to (23) in which m_0 is replaced by $m_0(1 - B\kappa)$, where B is a coefficient related to the unstretched, planar laminar-flame structure. The correction applies only to first order in κ (weak stretch).

Although analyses could be developed on the basis of the δ-function model of Chapter I, §10, activation-energy asymptotics, roughly like Chapter I, §9, are better justified and therefore preferable. For illustrative purposes, consider a one-reactant, one-step process in a gas with constant properties, λ, C_p, Le and Pr. In terms of the temperature T_0 of the unburnt gas, the adiabatic flame temperature T_f and the activation temperature T_a, parameters that arise are the nondimensional heat release $\alpha = (T_f - T_0)/T_0$ and the Zeldovich number $\beta = T_a(T_f - T_0)/T_f^2$. Use the adiabatic, planar, laminar-flame thickness $\delta = \lambda/MC_p$ and residence time $\rho_0\lambda/M^2C_p$ to nondimensionalize space coordinates and time, and let the fresh mixture move in the x direction into the stationary turbulent flame at velocity v_T. With $f = F/\delta$, $\xi = x/\delta - f$, $\eta = y/\delta$, $\zeta = z/\delta$, u, v and w representing the ξ, η and ζ components of velocity nondimensionalized by v_0, Y the ratio of the fuel mass fraction to its value in the fresh mixture, $\Theta = (T - T_0)/(T_f - T_0)$, $r = \rho/\rho_0 = [1 + \Theta\alpha/(1 - \alpha)]^{-1}$, and p the departure of the pressure from its initial value, divided by Ma^2, the conservation equations in the moving frame are

(26) $$\frac{\partial r}{\partial \tau} + \frac{\partial s}{\partial \xi} + \frac{\partial(rv)}{\partial \eta} + \frac{\partial(rw)}{\partial \zeta} = 0,$$

(27) $$r\frac{\partial u}{\partial \tau} + (s + \text{Pr}\,\nabla_t^2 f)\frac{\partial u}{\partial \xi} + rv\frac{\partial u}{\partial \eta} + rw\frac{\partial u}{\partial \zeta}$$
$$= -\frac{\partial p}{\partial \xi} + \text{Pr}\,\Delta' u + \tfrac{1}{3}\text{Pr}\,\frac{\partial}{\partial \xi}\left[\frac{\partial(s/r)}{\partial \xi} + \frac{\partial v}{\partial \eta} + \frac{\partial w}{\partial \zeta}\right],$$

(28) $$r\frac{\partial v}{\partial \tau} + (s + \text{Pr}\,\nabla_t^2 f)\frac{\partial v}{\partial \xi} + rv\frac{\partial v}{\partial \eta} + rw\frac{\partial v}{\partial \zeta}$$
$$= -\frac{\partial p}{\partial \eta} + f_\eta\frac{\partial p}{\partial \xi} + \text{Pr}\,\Delta' v$$
$$+ \tfrac{1}{3}\text{Pr}\left(\frac{\partial}{\partial \eta} - f_\eta\frac{\partial}{\partial \xi}\right)\left[\frac{\partial(s/r)}{\partial \xi} + \frac{\partial v}{\partial \eta} + \frac{\partial w}{\partial \zeta}\right],$$

a similar equation for w, and

$$r\frac{\partial \Theta}{\partial \tau} + (s + \nabla_t^2 f)\frac{\partial \Theta}{\partial \xi} + rv\frac{\partial \Theta}{\partial \eta} + rw\frac{\partial \Theta}{\partial \zeta} - \Delta'\Theta$$
$$= -r\frac{\partial Y}{\partial \tau} - (s + \text{Le}^{-1}\nabla_t^2 f)\frac{\partial Y}{\partial \xi} - rv\frac{\partial Y}{\partial \eta} - rw\frac{\partial Y}{\partial \zeta} + \text{Le}^{-1}\Delta' Y$$

(29) $$= \Lambda_L\beta^{n+1}Y^n\exp\{-\beta(1 + \Theta)/[1 - \alpha(1 - \Theta)]\}.$$

Here ∇_t is the transverse gradient in (η, ζ),

(30) $$s = r(u - f_\tau - vf_\eta - wf_\zeta),$$

(31) $$\Delta' = (1 + |\nabla_t f|^2)\frac{\partial^2}{\partial \xi^2} + \frac{\partial^2}{\partial \eta^2} + \frac{\partial^2}{\partial \zeta^2} - 2f_\eta\frac{\partial^2}{\partial \xi\partial \eta} - 2f_\zeta\frac{\partial^2}{\partial \xi\partial \zeta},$$

n is the reaction order, and $\Lambda_L = [2\Gamma(n + 1)]^{-1} + \cdots$, as an asymptotic expansion in β^{-1}.

With Le = $1 + l/\beta$ and l of order unity, an expansion for $\beta \to \infty$ gives as jump conditions [ ] across the reaction sheet (at $\xi = 0$),

$$[\Theta] = 0, \quad [Y] = 0, \quad [u] = 0, \quad [v] = 0,$$

$$\left[\frac{\partial \Theta^{(0)}}{\partial \xi}\right] = -\left[\frac{\partial Y^{(0)}}{\partial \xi}\right] = -(1 + |\nabla_t f|^2)^{-1/2} e^{\Theta^{(1)}/2},$$

(32) $\quad \left[\frac{\partial \Theta^{(1)}}{\partial \xi}\right] + \left[\frac{\partial Y^{(1)}}{\partial \xi}\right] = l\left[\frac{\partial Y^{(0)}}{\partial \xi}\right],$

$$[p] = \Pr(1 + |\nabla_t f|^2)\left[\frac{\partial u}{\partial \xi}\right] + \tfrac{1}{3}\Pr\left[\frac{\partial(s/r)}{\partial \xi}\right],$$

$$f_\eta\left[\frac{\partial u}{\partial \xi}\right] = -\left[\frac{\partial v}{\partial \xi}\right],$$

where $\Theta = \Theta^{(0)} + \beta^{-1}\Theta^{(1)} + \cdots$, $Y = Y^{(0)} + \beta^{-1}Y^{(1)} + \cdots$, and $\Theta^{(0)} = 1$, $Y^{(0)} = Y^{(1)} = 0$ for $\xi > 0$. Use of these enables (29) to be put to zero in analyzing $\xi \gtrsim 0$, of order unity or large.

In considering weak stretch, an expansion is introduced in a small parameter ϵ representing the magnitudes of gradients. Multiple-scale methods are useful for studying various aspects of flame dynamics, but for simplicity here consider instead matched asymptotic expansions in ϵ, with zones having ξ of order unity on each side of the reaction sheet matched to far-field zones where ξ is of order ϵ^{-1}. In the far fields zero Mach number equations for constant-density, viscous flows are obtained. In the near fields departures of the variables from their far-field upstream values may be expanded as

$$\Theta = \Theta_0(\xi) + \epsilon\Theta_1(\xi, \epsilon\eta, \epsilon\zeta, \epsilon\tau) + \cdots,$$

$$Y = Y_0(\xi) + \epsilon Y_1(\xi, \epsilon\eta, \epsilon\zeta, \epsilon\tau) + \cdots,$$

(33) $\quad s = s_0(\xi) + \epsilon s_1(\xi, \epsilon\eta, \epsilon\zeta, \epsilon\tau) + \cdots,$

$$u = u_0(\xi) + \epsilon u_1(\xi, \epsilon\eta, \epsilon\zeta, \epsilon\tau) + \cdots,$$

$$v = v_0(\xi) + \epsilon v_1(\xi, \epsilon\eta, \epsilon\zeta, \epsilon\tau) + \cdots,$$

$$p = p_0(\xi) + \epsilon p_1(\xi, \epsilon\eta, \epsilon\zeta, \epsilon\tau) + \cdots,$$

and with $f = \epsilon^{-1}f_0 + f_1 + \cdots$, differential equations for $\Theta_0, Y_0, \Theta_1, \cdots$ may be developed by substitution and collection of like powers of ϵ for describing the perturbed flame structure. When carried to second order and integrated with application of matching conditions, these equations eventually yield

$$(34) \quad B = \frac{1}{\alpha} \ln\left(\frac{1}{1-\alpha}\right) + \frac{l}{2}\left(\frac{1-\alpha}{\alpha}\right) \int_0^{\alpha/(1-\alpha)} x^{-1} \ln(1+x)\, dx$$

for the previously cited coefficient of κ in the correction to m_0. Different formulas for B are obtained from theories having different assumptions about property variations [53]. Many aspects of flame dynamics in turbulent fields are obtainable from the results [9], [30], [40], [45].

Influences of moderate and strong stretch. Analyses of planar laminar flames have considered moderate stretch (κ of order unity) and strong stretch (κ large), showing, for example, that under appropriate conditions abrupt disruption of the flame sheet (laminar flame extinction) may occur [45], [50], [54]–[62]. These results provide underlying information having relevance to structures of premixed turbulent flames at values of l_k/δ less than infinity.

If wrinkled flame extinctions occur, then the reaction-sheet picture, as developed thus far, cannot tell the whole story about premixed turbulent flames. Even (22) no longer seems true because some of the fuel escapes unburnt. Complications of this kind are expected in traveling from the reaction-sheet regime toward the stirred reactor regime. It has been indicated above that some investigators have suggested turbulent-flame extinction if the turbulence scales become too small. Alternatively, the previously mentioned abrupt transitions to different modes of combustion may occur. Analyses of turbulent flames involving wrinkled laminar flame sheets with holes and analyses of the dynamics of holes in flame sheets could help to clarify the situation.

5. Prospects. Because of intensive efforts currently devoted to theories of turbulent combustion by a wide variety of approaches, prospects for future advances are bright. Nevertheless, there are so many different avenues to the subject that there is room for many more researchers.

Of the regimes that have been identified here, the greatest amount of recent progress has been achieved in those involving reaction sheets, both for diffusion flames and premixed flames. The intensity of work on premixed flames with reaction sheets is high; significant further advances in understanding of this topic may be expected in the future. Especially important is further study of high-intensity limits in the large-scale regime. For diffusion flames with reaction sheets there appears now to be somewhat less activity; since the outstanding problems here (e.g., the need for further perturbation studies of turbulent-flow effects of finite-thickness sheet structure) are becoming more difficult, less impressive progress might be obtained. The main outstanding needs for basic understanding appear to lie in regimes of distributed reaction and in better identification of locations and characters of boundaries between regimes. Knowledge of underlying structures about which perturbations might be developed is entirely lacking for distributed-reaction regimes; the intensity of effort is low here, and without more involvement, prospects for progress are dim.

From the discussions that have been given it will be seen that perturbation methods have played an important role in the progress that has been achieved and are likely to continue to do so. Moment methods have a well defined position; the best road ahead for them would appear to be one of slow and careful evolution, with close attention paid to variable-density effects for conserved scalars. Approximating probability-density functions for conserved scalars also appears to hold prospects for further slow evolution, but for scalars with sources their prospects are decidedly dim. Approaches in which evolutions of probability-density functions are calculated have now emerged as tools usable for engineering problems that are not too complicated, but potential sources of error lie at the closure level for these techniques; most of the current effort here appears to be directed toward the applications instead of being focused on the underpinnings where there are important and challenging problems begging for further study. The development of methods for the direct numerical integration of the conservation equations may safely be predicted to continue to attract diligent researchers who will make progress at a rate commensurate with rates of advance of electronic computers. Theories of turbulent combustion based on the full fundamental conservation equations while building from concepts of coherent structures or of strange attractors are distant hopes.

REFERENCES

[1] F. A. WILLIAMS, *Theory of combustion in laminar flows*, Ann. Rev. Fluid Mech., 3 (1971), pp. 171–188.
[2] P. A. LIBBY AND F. A. WILLIAMS, eds., *Turbulent Reacting Flows*, Springer-Verlag, Berlin, 1980.
[3] N. PETERS AND F. A. WILLIAMS, *Coherent structures in turbulent combustion*, in The Role of Coherent Structures in Modelling Turbulence and Mixing, J. Jimenez, ed., Springer-Verlag, Berlin, 1981, pp. 364–393.
[4] W. P. JONES AND J. H. WHITELAW, *Calculation methods for reacting turbulent flows: a review*, Combustion and Flame, 48 (1982), pp. 1–26.
[5] H. EICKHOFF, *Turbulent hydrocarbon jet diffusion flames*, Prog. Energy Combust. Sci., 8 (1982), pp. 159–169.
[6] W. C. STRAHLE, *Duality, dilatation, diffusion and dissipation in reacting turbulent flows*, Nineteenth Symposium (International) on Combustion, The Combustion Institute, Pittsburgh, 1982, pp. 337–347.
[7] R. GUNTHER, *Turbulence properties of flames and their measurement*, Prog. Energy Combust. Sci., 9 (1983), pp. 105–154.
[8] G. I. SIVASHINSKY, *Instabilities, pattern formation, and turbulence in flames*, Ann. Rev. Fluid Mech., 15 (1983), pp. 179–199.
[9] P. CLAVIN, *Dynamical behavior of premixed flame fronts in laminar and turbulent flows*, Prog. Energy Combust. Sci., to appear, 1985.

[10] S. B. POPE, *PDF methods for turbulent reactive flows*, Prog. Energy Combust. Sci., to appear, 1985.
[11] F. A. WILLIAMS, *Asymptotic methods in turbulent combustion*, AIAA Preprint 84-0475, 1984.
[12] K. SAGARA AND S. TSUGÉ, *A bimodal maxwellian distribution as the equilibrium solution of the two-particle regime*, Phys. Fluids, 25 (1982), pp. 1970–1977.
[13] L. P. KADANOFF, *Roads to chaos*, Physics Today, 36 (1983), pp. 46–53.
[14] E. HOPF, *Statistical hydromechanics and functional calculus*, J. Rat. Mech. Anal., 1 (1952), pp. 87–124.
[15] T. S. LUNDGREN, *Distribution functions in the statistical theory of turbulence*, Phys. Fluids, 10 (1967), pp. 969–975.
[16] A. PAPOULIS, *Probability, Random Variables and Stochastic Processes*, McGraw-Hill, New York, 1965.
[17] H. R. BAUM, R. G. REHM, P. D. BARNETT AND D. G. CORLEY, *Finite difference calculations of buoyant convection in an enclosure, I. the basic algorithm*, SIAM J. Sci. Stat. Comp., 4 (1983), pp. 117–135.
[18] A. F. GHONEIM, A. J. CHORIN AND A. K. OPPENHEIM, *Numerical modelling of turbulent combustion in premixed gases*, Eighteenth Symposium (International) on Combustion, The Combustion Institute, Pittsburgh, 1981, pp. 1375–1383.
[19] S. N. B. MURTHY, ed., *Turbulent Mixing in Nonreactive and Reactive Flows*, Plenum, New York, 1975.
[20] P. BRADSHAW, ed. *Turbulence*, Springer-Verlag, Berlin, 1978.
[21] P. CLAVIN AND F. A. WILLIAMS, *Theory of premixed-flame propagation in large-scale turbulence*, J. Fluid Mech., 90 (1979), pp. 589–604.
[22] J. B. MOSS, *Simultaneous measurements of concentration and velocity in an open turbulent flame*, Combustion Sci. Tech., 22 (1980), pp. 119–129.
[23] P. A. LIBBY AND K. N. C. BRAY, *Countergradient diffusion in premixed turbulent flames*, AIAA J., 19 (1981), pp. 205–213.
[24] I. G. SHEPHERD, J. B. MOSS AND K. N. C. BRAY, *Turbulent transport in a confined premixed flame*, Nineteenth Symposium (International) on Combustion, The Combustion Institute, Pittsburgh, 1982, pp. 423–431.
[25] D. T. PRATT, *Mixing and chemical reaction in continuous combustion*, Prog. Energy Combust. Sci., 1 (1976), pp. 73–86.
[26] D. B. SPALDING, *Mathematical models of turbulent flames: A review*, Combustion Sci. Tech., 13 (1976), pp. 3–25.
[27] A. S. C. MA, D. B. SPALDING AND R. L. T. SUN, *Application of "ESCIMO" to the turbulent hydrogen-air diffusion flame*, Nineteenth Symposium (International) on Combustion, The Combustion Institute, Pittsburgh, 1982, pp. 393–402.
[28] F. A. WILLIAMS, *Current problems in combustion research*, in Dynamics and Modelling of Reactive Systems, W. E. Stewart, W. H. Ray and C. C. Conley, eds., Academic Press, New York, 1980, pp. 293–314.
[29] G. DAMKÖHLER, *Einflüsse de Strömung, Diffusion und des Wärmeüberganges auf die Leistung von Reaktionsöfen*, Z. Elektrochem., 42 (1936), pp. 846–862.
[30] F. A. WILLIAMS, *Combustion Theory*, second edition, Benjamin-Cummings, Menlo Park, CA, 1985.
[31] A. S. MONIN AND A. M. YAGLOM, *Statistical Fluid Mechanics*, MIT Press, Cambridge, MA, Vol. 1, 1971, Vol. 2, 1975.
[32] N. PETERS, *Asymptotic analysis of nitric oxide formation in turbulent diffusion flames*, Combustion Sci. Tech., 19 (1978), pp. 39–49.
[33] P. A. LIBBY AND F. A. WILLIAMS, *Some implications of recent theoretical studies in turbulent combustion*, AIAA J., 19 (1981), pp. 261–274.

[34] A. LIÑÁN, *The asymptotic structure of counterflow diffusion flames for large activation energies*, Acta Astronautica, 1 (1979), pp. 1007–1039.
[35] F. A. WILLIAMS, *A review of flame extinction*, Fire Safety J., 3 (1981), pp. 163–175.
[36] N. PETERS, *Local quenching due to flame stretch and nonpremixed combustion*, Combustion Sci. Tech., 30 (1983), pp. 1–17.
[37] N. PETERS AND F. A. WILLIAMS, *Liftoff characteristics of turbulent jet diffusion flames*, AIAA J., 21 (1983), pp. 423–429.
[38] J. JANICKA AND N. PETERS, *Prediction of turbulent jet diffusion flame lift-off using a PDF transport equation*, Nineteenth Symposium (International) on Combustion, The Combustion Institute, Pittsburgh, 1982, pp. 367–374.
[39] A. M. KLIMOV, *Flame propagation under conditions of strong turbulence*, Dokl. Akad. Nauk SSSR, 221 (1975), pp. 56–59. (In Russian.)
[40] P. CLAVIN AND F. A. WILLIAMS, *Effects of molecular diffusion and of thermal expansion on the structure and dynamics of premixed flames in turbulent flows of large scale and low intensity*, J. Fluid Mech., 116 (1982), pp. 251–282.
[41] B. KARLOVITZ, D. W. DENNISTON, JR., D. H. KNAPSCHAEFER, AND F. H. WELLS, *Studies on turbulent flames*, Fourth Symposium (International) on Combustion, Williams and Wilkins, Baltimore, 1953, pp. 613–620.
[42] A. M. KLIMOV, *Laminar flame in a turbulent flow*, Zh. Prikl. Mekh. Tekh., 3 (1963), pp. 49–58. (In Russian.)
[43] F. A. WILLIAMS, *A review of some theoretical considerations of turbulent flame structure*, in Analytical Numerical Methods for Investigation of Flow Fields with Chemical Reactions, Especially Related to Combustion, M. Barrère, ed., AGARD Conference Proceedings No. 164, AGARD, Paris, 1975, pp. III-1 to III-25.
[44] J. D. BUCKMASTER, *The quenching of two-dimensional premixed flames*, Acta Astronautica, 6 (1979), pp. 741–769.
[45] P. CLAVIN AND G. JOULIN, *Premixed flames in large scale and high intensity turbulent flow*, J. de Physique-Lettres, 44 (1983), pp. L-1 to L-12.
[46] M. MATALON, *On flame stretch*, Combustion Sci. Tech., 31 (1983), pp. 169–181.
[47] S. H. CHANG AND C. K. LAW, *An invariant derivation of flame stretch*, Combustion and Flame, 55 (1984), pp. 123–125.
[48] G. I. SIVASHINSKY, *Nonlinear analysis of hydrodynamic instability in laminar flames—I. Derivation of basic equations*, Acta Astronautica, 4 (1977), pp. 1177–1206.
[49] P. CLAVIN AND F. A. WILLIAMS, *Effects of Lewis number on propagation of wrinkled flames in turbulent flow*, in Combustion in Reactive Systems, Vol. 76 of Progress in Astronautics and Aeronautics, J. R. Bowen, N. Manson, A. K. Oppenheim and R. I. Soloukhin, eds., American Institute of Aeronautics and Astronautics, New York, 1981, pp. 403–411.
[50] J. D. BUCKMASTER AND G. S. S. LUDFORD, *Theory of Laminar Flames*, Cambridge Univ. Press, Cambridge, 1982.
[51] P. PELCÉ AND P. CLAVIN, *Influence of hydrodynamics and diffusion upon the stability limits of laminar premixed flames*, J. Fluid Mech., 124 (1982), pp. 219–237.
[52] M. MATALON AND B. J. MATKOWSKY, *Flames as gasdynamic discontinuities*, J. Fluid Mech., 124 (1982), pp. 239–259.
[53] P. CLAVIN AND P. L. GARCÍA-YBARRA, *The influence of the temperature dependence of diffusivities on the dynamics of flame fronts*, J. de Mécanique Théorique et Appliqué, 2 (1983), pp. 245–263.
[54] V. M. GREMYACHKIN AND A. G. ISTRATOV, *On a steady flame in a stream with a velocity gradient*, Gorenie i Vzriva, Nauka, Moscow, 1972, pp. 305–308. (In Russian.)
[55] G. I. SIVASHINSKY, *On a distorted flame as a hydrodynamic discontinuity*, Acta Astronautica, 3 (1976), pp. 889–918.

[56] J. D. BUCKMASTER, *The quenching of a deflagration wave held in front of a bluff body*, Seventeenth Symposium (International) on Combustion, The Combustion Institute, Pittsburgh, 1979, pp. 835–842.

[57] Y. B. ZEL'DOVICH, G. I. BARENBLATT, V. B. LIBROVICH, AND G. M. MAKHVILADZE, *The Mathematical Theory of Combustion and Explosion*, Nauka, Moscow, 1980, pp. 272–277. (In Russian.)

[58] P. A. LIBBY AND F. A. WILLIAMS, *Structure of laminar flamelets in premixed turbulent flames*, Combustion and Flame, 44 (1982), pp. 287–303.

[59] J. D. BUCKMASTER AND D. MIKOLAITIS, *The premixed flame in a counterflow*, Combustion and Flame, 47 (1982), pp. 191–204.

[60] P. A. LIBBY AND F. A. WILLIAMS, *Strained premixed laminar flames under nonadiabatic conditions*, Combustion Sci. Tech., 31 (1983), pp. 1–42.

[61] K. SESHADRI AND N. PETERS, *The influence of stretch on a premixed flame with two-step kinetics*, Combustion Sci. Tech., 33 (1983), pp. 35–63.

[62] P. A. LIBBY, A. LIÑÁN, AND F. A. WILLIAMS, *Strained premixed laminar flames with nonunity Lewis numbers*, Combustion Sci. Tech., 34 (1983), pp. 257–293.

CHAPTER IV

Detonation in Miniature

WILDON FICKETT

Introduction. The advantages of studying simple model equations, such as the Burgers equation or the Korteweg–de Vries equation, are well known: the essential properties of the original are more clearly revealed by the stripping away of details which tend to obscure the view, and investigation is greatly facilitated by the model's simplicity. Following an introduction and a brief review of the current status of detonation theory, we show here what can be done with a set of model equations, or *mathematical analog*, for compressible flow with chemical reaction.

Our topic is detonation. An explosive is a material capable of violent exothermic chemical reaction, but normally resting quietly in a state of metastable chemical equilibrium. It is "stable" under ambient conditions only because the reaction rate is so slow. A local stimulus can initiate a reaction wave which spreads through the material. This wave can be one of two types: a deflagration or a detonation. A deflagration (flame) is a slow (far subsonic) wave. Transport processes—viscosity, heat conduction, and matter diffusion—are dominant. Changes in momentum and kinetic energy are small; the pressure change through the wave can, to a good approximation, be neglected. A detonation is at the other extreme. It runs at supersonic speed— six- to eight thousand m/s in liquids and solids. Changes in momentum and kinetic energy dominate: compressibility and inertia are the important properties. The transport processes, all-important in the flame, are relatively unimportant here. The leading element of the wave is a shock, and the high temperature and pressure produced by it are maintained by inertial confinement—the outer layers of the material are not appreciably displaced before the reaction in the center is over. The high wave speed leads to very high rates of energy conversion (chemical-bond to heat): a ten-square-centimeter section of detonation front in a solid explosive like TNT converts energy at a rate of 10^{11} W, a figure comparable to the total electric generating capacity of the United States.

Section 1 covers the physical system—the governing equations, an introduction to detonation theory, and a brief overview of theoretical work in the field. The remainder of our presentation is devoted to the mathematical analog and what can be done with it, but we do present results for the physical system at some points for comparison. In §2 we describe the analog and show how it can be "derived" from the physical system. In §3 we give an overview of a number of successful applications of it. For two particular applications we give a more detailed presentation. The groundwork for these is given in §4, which lays out some of the basics which will be needed. The two applications follow: the hydrodynamic stability of the steady detonation in §5, and the initiation of detonation by a shock wave in §6. The Notes at the end of the chapter contain background references and remarks for each section.

1. Physical system. The governing equations usually taken to describe the system are the so-called Euler equations of inviscid compressible flow, with chemical reaction added. These are obtained from the compressible Navier–Stokes equations by dropping the transport terms (viscosity, heat conduction, and diffusion). We refer to these equations as the *physical system*, or the *original*, for short, as opposed to the *analog* which models them. This section is devoted to the physical system. We set down the governing equations, give a short introduction to detonation theory, and conclude with a brief overview of theoretical work in the field. Several of the figures of this section will also serve to illustrate the similar results for the analog.

1.1. Governing equations. We specialize to one-dimensional flow throughout. An idealized version of the experimental situation is shown in Fig. 1.1. The explosive is confined in a semi-infinite tube with rigid, perfectly slippery walls, bounded at the left by a movable piston. The initial condition is the unreacted spatially uniform explosive. The piston is under the control of the experimenter. The boundary condition is the specification of the piston

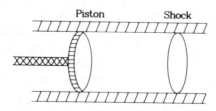

FIG. 1.1. *The piston problem: one-dimensional flow in a rigid-walled tube with prescribed piston motion.*

(boundary) velocity u_b as a function of time. We take the piston to be placed initially at $x = 0$, and motionless for $t < 0$.

In a real system we have edge effects, and often only limited control of the piston. For a gaseous explosive we have boundary layers at the walls, and the "piston" may be just the rigid end of the tube or possibly the driver section of a shock tube. For a solid explosive, no material is strong enough to serve as a perfectly rigid wall. The "piston" may be the projectile of a compressed-gas gun, or, for higher pressure work, a projectile driven by another piece of explosive.

The governing equations are the usual Euler equations of inviscid compressible flow, with chemical reaction added. For the most part, we include only a single chemical reaction, represented schematically by

$$A \rightarrow B.$$

We describe the progress of the chemical reaction by a *progress variable* λ, which goes from zero to one as the reaction proceeds from pure A to pure B. Thus λ is the mass fraction of B, and defines the chemical composition of the system. Under the assumptions each fluid element, or *particle*, is a closed adiabatic system. It is thus convenient to use the material derivative (total time derivative for a *particle*) defined by

$$\dot{f} = f_t + u f_x,$$

where the dot denotes the material derivative, t and x are time and distance, and u is the particle velocity. The Euler equations are

(1.1a) $$\dot{\rho} + \rho u_x = 0,$$

(1.1b) $$\rho \dot{u} + p_x = 0,$$

(1.1c) $$\dot{e} + p \dot{v} = 0,$$

(1.1d) $$\dot{\lambda} = r,$$

(1.1e) equation of state $p = p(\rho, e, \lambda),$

(1.1f) reaction rate $r = r(\rho, e, \lambda),$

with density ρ, pressure p, internal energy e, and specific volume $v = 1/\rho$. The dependent variables are (ρ, e, λ, u). The pressure p and reaction rate r are given state functions describing the material.

The temperature does not appear in these equations; it is not needed because the transport properties are neglected. Most experiments involve mechanical measurements of pressure or particle velocity. From these, density, energy, and λ (under certain assumptions) can often be inferred by application of the conservation conditions. So our set of variables is the convenient one for most applications. If a temperature is needed, as, for example, if one were to assume an Arrhenius form for the reaction rate, then of course one would also need the "thermal" equation of state $T = T(\rho, e, \lambda)$.

A commonly used form for the equation of state is the simple polytropic gas form

$$p = (\gamma - 1)\rho(e + q\lambda),$$

with constants q and γ; q is the heat of reaction and γ is the negative logarithmic slope of the isentrope, $\gamma = -(\partial \ln p/\partial \ln v)_S$. For gaseous explosives, it is a good approximation to describe the system as an ideal gas with constant specific heat. This gives the above form with γ the usual ratio of specific heats. For solid explosives, the same form with $\gamma \approx 3$ gives a fairly realistic description. Its simplicity makes it attractive for theoretical work.

The equations (1.1) are hyperbolic, with three families of characteristics, which we denote by $(+)$, $(-)$, and (0). The forward $(+)$ and backward $(-)$ acoustic families have velocities $u + c$ and $u - c$, respectively, and the particle (0) family has velocity u. Here c is the so-called frozen sound speed, defined in terms of the partial derivatives of the equation of state by

$$c^2 = p_\rho - \frac{p}{\rho^2} p_e .$$

All thermodynamic variables other than the composition are assumed to be in local thermodynamic equilibrium. The only entropy-producing processes are the chemical reaction and passage through a shock.

In this formulation shocks are mathematical discontinuities. The simplest case is the flat-topped shock, Fig. 1.2, generated by a constant-velocity piston moving into a nonreactive material. The shock is described by the Rankine–Hugoniot relations—algebraic relations expressing the conservation of mass, momentum, and energy across it. A common form is

(1.2a) $$u^2 = (p - p_i)(v_i - v),$$

(1.2b) $$\rho_i^2 D^2 = (p - p_i)/(v_i - v),$$

(1.2c) $$e - e_i = \tfrac{1}{2}(p + p_i)(v_i - v),$$

(1.2d) $$\lambda = \lambda_i.$$

Here D is the shock velocity, and subscript i and no subscript denote the state just before and just behind the shock (we take $u_i = 0$). We will use the term *shock state* for the state just behind the shock, and a similar terminology for individual quantities, as, for example, the *shock density*. The last equation simply states that no reaction takes place across the instantaneous shock jump. The first two equations are independent of the material,

DETONATION IN MINIATURE

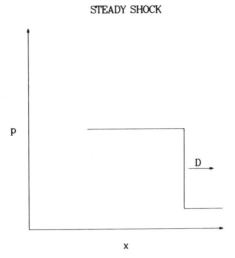

FIG. 1.2. *Steady flat-topped shock produced by a constant-velocity piston.*

but the third depends on the equation of state. In the p-v plane, Fig. 1.3, the second equation (1.2b) describes the *Rayleigh line*, a straight line passing through the initial state (p_i, v_i), point I in the figure. Its slope is proportional to the square of the shock velocity. The third equation (1.2c) is the shock Hugoniot, the locus of all shock states which can be reached from the given

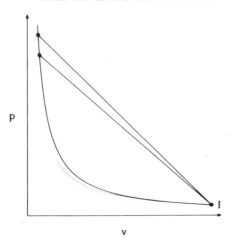

FIG. 1.3. *State curves and Rayleigh lines for the physical system. Shock Hugoniot through the initial state (solid curve) and isentrope through the same point (dashed curve).*

initial state. At any point on the *shock Hugoniot*, the slope of the Rayleigh line through that point gives the shock velocity, and the first equation (1.2a) gives the shock particle velocity (particle velocity just behind the shock). The isentrope through the initial state, the dashed curve in the figure, lies below the shock Hugoniot.

Signals overtaking the shock from behind are reflected back into the following flow. The shock acts as a floating boundary, with the Rankine–Hugoniot relations (1.2) applying as a boundary condition along its path.

1.2. Detonation theory. The so-called ZND model of a detonation wave is shown in Fig. 1.4. The leading element of the wave is a shock. Passage through the shock instantaneously places each incoming particle in a brand-new environment: a highly compressed state at a temperature of several thousand degrees. The rapid reaction which ensues goes very quickly to completion in a thin *reaction zone* which follows the shock. Behind the reaction zone is some *following flow*, most commonly a rarefaction wave. The energy released by the chemical reaction propels the shock at a high speed and pressure. The shock and its associated reaction zone, often referred to as the detonation front, moves rapidly away from the left, or *rear*, boundary where the detonation was initiated. The following rarefaction wave connects, or *matches*, the high pressure and (forward) particle velocity at the end of the reaction zone to the much smaller values at the rear boundary.

The simplest case is the steady detonation, in which the shock velocity is constant and the flow in the reaction zone is steady in a frame attached to the shock. (A following rarefaction wave is necessarily nonsteady.) To

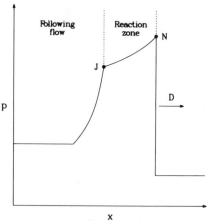

FIG. 1.4. *Unsupported* (CJ) *detonation.*

find steady solutions, we transform the governing equations to a frame attached to the shock and set time derivatives to zero. The resulting equations have solutions for all values of D above a certain minimum value. In a manner to be explained, specifying a particular rear boundary condition picks out one member of this one-parameter family of solutions.

For the steady flow (within the reaction zone), the first three governing equations (1.1a–c) reduce to the three algebraic shock relations (1.2a–c), except that the plain symbols now refer to any point within the reaction zone, and the energy e now depends on λ as well as on p and ρ. As we shall see in more detail below, these equations give the state as a function of λ through the reaction zone for a given value of D. This result, which is independent of the reaction rate, gives most of the information we usually need to know. To get the dependence on x, we must be given the reaction rate (1.1f). The rate equation (1.1d) then becomes an ODE for $\lambda(x)$:

$$\frac{d\lambda}{dx} = -\frac{r(p(\lambda), \rho(\lambda), \lambda)}{D - u(\lambda)},$$

$$\lambda = 0 \quad \text{at } x = 0,$$

where $x = 0$ at the shock, and $p(\lambda)$, $\rho(\lambda)$ and $u(\lambda)$ are obtained from the solution of the algebraic relations (1.2a–c).

As we did for the shock in Fig. 1.3, we diagram the Rayleigh line (1.2b) and Hugoniot (1.2c) relations in the p–v plane in Fig. 1.5. With the inclusion of reaction, the Hugoniot equation (1.2c) takes on a new meaning. Writing it with the dependence of e on λ explicitly indicated,

$$e(p, v, \lambda) - e(p_i, v_i, \lambda = 0) = \tfrac{1}{2}(p + p_i)(v_i - v),$$

we see that it describes a one-parameter (λ) family of curves, the *partial-reaction Hugoniot* curves. As λ increases from zero, the heat of reaction displaces the entire curve upward. Of special interest are the two extremes shown in the figure: the no-reaction ($\lambda = 0$) curve, which represents an inert shock in the reactants, and the complete-reaction ($\lambda = 1$) curve, the *detonation Hugoniot* curve, on which the state at the end of the reaction zone must lie.

For a given shock velocity D, all states must lie on the corresponding Rayleigh line, and since the Hugoniot relation must also be satisfied, they are confined to the segment of the Rayleigh line bounded by its intersection with the $\lambda = 0$ and $\lambda = 1$ curves, shown as the solid lines in the figure. Let us trace the state history of a particle as it traverses the steady solution. It starts at the initial state I. As it passes through the lead shock, its state jumps to a point N on the $\lambda = 0$ curve. As it moves through the reaction zone, reaction proceeds and λ increases. At each value of λ, it must lie on the corresponding partial reaction curve. Thus its state slides down the solid portion of the Rayleigh line as indicted by the arrow, crossing partial-reaction Hugoniot curves of increasing λ as it goes. At the end of the reaction zone it has reached the $\lambda = 1$ curve at point S or J[1].

The lowest possible Rayleigh line is the one that is tangent to the $\lambda = 1$ curve, the lower Rayleigh line in the figure. The corresponding detonation velocity is the minimum steady value consistent with the conservation conditions. The corresponding detonation is called the *Chapman–Jouguet* or CJ detonation, and point J is called the CJ point. Physically, as we shall see shortly, this is an *unsupported detonation* (that is, one that is not pushed from behind). Higher Rayleigh lines represent *supported*, or *overdriven* detonations, forced to run at a higher velocity by being pushed from behind by the piston. A complete-reaction point like S is called a strong point. An important property is that the flow at a strong point is subsonic like that behind an inert shock (that is, $u + c > D$) and the flow at the CJ point J is sonic ($u + c = D$). Thus an overdriven detonation is subject to degradation by a rarefaction wave overtaking it from behind, while the CJ detonation is not. (In the CJ case the head of a following rarefaction wave, running at the $(+)$ characteristic velocity $u + c$, travels with precisely the same speed as the end of the reaction zone.)

A particular steady solution can be generated by appropriate choice of the rear boundary condition. Suppose we prescribe a constant boundary velocity u_b (that is, move the piston, Fig. 1.1, at this constant velocity). Which member of the family of possible steady solutions (for different values of D) described above will be produced by a given u_b? Let subscript 1 denote the state at the end of the reaction zone, with u_1 the particle velocity there. Let the value of u_1 at the CJ (tangent) point be u_j. For the discussion, start with $u_b > u_j$ and consider how the steady solution changes as u_b decreases. For $u_b > u_j$ we have an overdriven detonation, as shown in Fig. 1.6. The particular one is that terminating at the particular strong point at which $u_1 = u_b$. The following flow is a constant state with $u = u_1 = u_b$, $p = p_1$, $v = v_1$, etc. As u_b is decreased toward u_j, and at $u_b = u_j$, we have the same shape profile at successively lower detonation velocities and pressures, as point S moves down toward point J, and point N drops correspondingly. If now u_b is decreased below u_j, the detonation velocity and reaction zone remain unchanged, for the Rayleigh line can drop no lower. The "slack" is taken up by the replacement of the constant following flow by a rarefaction wave (Fig. 1.4) which matches the CJ particle velocity u_j to the lower piston velocity. The smaller u_b, the larger the amplitude of this rarefaction. Thus for $u_b < u_j$ the detonation is "unsupported", and runs at the minimum (CJ) velocity, regardless of what is happening behind.

---

[1] The point labels N, J, and S stand for the von Neumann point, the Chapman–Jouguet point, and a strong point, respectively.

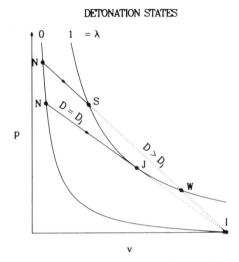

Fig. 1.5. *Detonation states in the p–v plane.*

For the simple one-reaction system we have used so far, the above discussion exhausts the possibilities for the steady solution. For more complicated systems (for example, those with more than one reaction) a different type of unsupported steady solution is possible, the so-called *eigenvalue detonation*. It runs faster than the CJ detonation, so that its Rayleigh line is like the upper one of Fig. 1.5, but it ends at a *weak point*, such as point

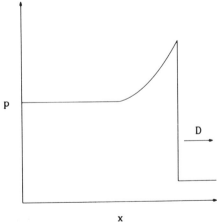

Fig. 1.6. *Supported (overdriven) detonation.*

W in the figure, on the $\lambda = 1$ curve *below* the CJ point, instead of at the strong point S. To see how it gets there takes more explanation than we have space for here. Some examples of eigenvalue detonations will be alluded to in §3.

1.3. Current work. We review here some of the recent theoretical work on detonation. We group the work under four topics: (1) perturbations on the steady solution, (2) geometrical acoustics, (3) initiation of detonation, and (4) granularity of the explosive. References to some of the older work are given in the Notes at the end of the chapter.

Perturbations on the steady solution. The one-dimensional steady solution is an idealization. In practice, perturbations on it must be considered. In a charge of finite length, there may not be time for attainment of the steady state. Or the system may be displaced from the steady state, once achieved, by a perturbation. Or the state may be steady, but not quite one-dimensional. We divide the work on perturbations of the steady solution into two headings: (1) nearly steady one-dimensional flow (including stability of the steady solution), and (2) nearly one-dimensional steady flow.

For time-dependent displacements from the steady state, the most common approach has been the usual stability analysis. One considers small (in general three-dimensional) perturbations on the one-dimensional steady solution and asks whether these grow or decay. In the process one discovers the natural frequencies and growth rates of the system (poles of the transfer function). A good deal of work has been done in past years along these lines. The results suggest that the steady solution is likely to be unstable. Experimentally, most, if not all, real detonations appear to fall into this category. The typical manifestation is the presence of transverse waves on the front, made up of triple-shock (Mach) configurations. If not confined by tube walls, (as in a spherical detonation, for example) these propagate randomly in all directions, giving the front a fine-scale, random, time-dependent cellular structure. In a tube, they may couple to the tube's acoustic modes to give beautiful regular patterns.

Another approach to the stability problem is to consider the long-wavelength limit under the rather drastic assumption that reaction is complete within the shock, so that the entire reaction zone and lead shock are treated as a single discontinuity. The stability of the system then depends entirely on the shape of the detonation Hugoniot ($\lambda = 1$) curve. Following this line of attack, Fowles [15] has studied the "stimulated and spontaneous emission" of acoustic waves from the front. He considers an acoustic wave obliquely incident on the front (from behind) and inquires into the nature of the reflected wave. The reflection coefficient can become infinite ("spontaneous emission") and thus represents a neutral stability boundary. Majda and Rosales [22] have extended this approach into the nonlinear regime, obtaining structures resembling the observed transverse waves, and a non-linear equation which describes their propagation on the front.

J. D. Bdzil of this laboratory has applied modern perturbation techniques to a number of detonation problems. Bdzil [2] studied one-dimensional time-dependent flow in a model two-reaction system. The first reaction, which releases most of the energy, is instantaneous; the second reaction, which releases the rest, has a finite rate. If the fraction of the energy released by the second reaction is δ^2, then at order δ "the evolution of the detonation proceeds as if it were a simple wave with independent variables x and δt." One obtains a relatively simple equation describing the approach to the steady solution.

To the author's knowledge, the stability properties of the eigenvalue type of steady solution have not been studied. This could be an interesting topic, since the properties of the steady eigenvalue solution are quite different from those of the normal (CJ) type.

Turning now to slightly non-one-dimensional steady solutions, the standard problem is the two-dimensional steady detonation in a confining tube. The main edge effect (in condensed explosives) is the outward displacement of the tube walls by the high pressure of the detonation. Using a perturbation approach, with the deflection angle of the wall as the small parameter, Bdzil [3] has solved this problem to a good approximation. Engelke and Bdzil [8] used this solution to deduce the approximate reaction rate in nitromethane from measurements of the wave curvature and the change in detonation velocity with tube diameter.

Geometrical acoustics. We turn now to the topic of geometrical acoustics in reaction flow, and what we shall call "geometrical detonation dynamics". These problems have apparently not been studied; it would seen that they ought to receive some attention. The geometrical acoustics problem is "How do acoustic wave fronts propagate in a given reactive flow field?" We may think, for example, of the (three-dimensional) propagation of acoustic wave fronts in a steady detonation reaction zone. In a nonreactive flow, the wave front propagates at the local sound speed and the energy propagates along the rays. In a reactive flow, it is clear that the wave front propagates, at least formally, with the frozen speed, but it is not clear how the energy propagates. Consideration of the way in which a one-dimensional rarefaction wave propagates into a reactive material in chemical equilibrium leads one to suspect that the equilibrium sound speed will be important here.

By "geometrical detonation dynamics" we mean the problem for detonation waves corresponding to that which Whitham [27, Chap. 8], has treated so successfully for shock waves, which he terms "geometrical shock dynamics". The problem is to construct an approximate theory for the way in which a detonation front propagates under conditions of varying geometry, as, for example, how it goes around a corner. Looking at Bdzil's results for the steady two-dimensional detonation, one is led to conjecture that the result would contain a local eigenvalue problem for the velocity of each element of the front. But the flow is subsonic at the shock, so the influence of neighboring elements will have to enter in some way. A "geometrical detonation dynamics" theory would probably be useful in numerical hydrodynamics codes. The reaction zone is so thin (≈ 30 μm in a good condensed explosive) that it is difficult to treat numerically. An attractive possibility for some purposes would be to treat the front as a discontinuity in the main calculation, pausing periodically to update the velocities of its elements from a side calculation based on the "geometrical detonation dynamics" theory.

Initiation of detonation. Our next topic is the initiation of detonation, both intentional and accidental. The two principal mechanisms are (1) initiation by a shock wave, and (2) an initial deflagration which turns into a detonation. Shock initiation, a common accidental mechanism, is also the principal means of intentionally starting a detonation. A typical detonator consists of a small charge of more sensitive explosive, itself initiated by an electrically exploded wire. Initiation of the main charge is by means of the shock wave sent into it by the detonator. For some applications this process must be extremely reliable and well controlled, and so it has been extensively studied. The standard problem for fundamental study is the initiation of detonation by a plane shock wave. Experimentally, the initiating shock wave is usually generated in a precise and controlled manner by the collision of a compressed-gas gun projectile with the explosive being studied. Instrumentation of the explosive with pressure gauges, plus appropriate analysis, yields information about the local energy release rate, so experiments of this type are of fundamental as well as applied interest.

The pressure of the initiating shock is small compared to the ultimate detonation pressure, so the experiment spans a wide range of conditions, which tends to make analytic treatment difficult. Similarity solutions which bear some resemblance to the real flows have been constructed by Cowperthwaite [6], (see also Holm and Logan [17]), but apart from this, little has been done. Fickett [12], [13] has treated the problem in the limiting case of small heat release. This work is reviewed in §6. Although this limiting case has some interest in its own right, it is rather far removed from the standard experiment. This appears to be a difficult area for analytic work. Possibly some sort of "adiabatic" or quasi-steady treatment of the reaction zone for the case of a rate with mild temperature dependence would be feasible.

The other initiation mechanism, the deflagration-to-detonation transition (DDT) is usually associated with accidental initiation of detonation. Thus, for example, using a more energetic rocket propellant increases the chance of DDT and consequent destruction of the rocket. A great deal of computational and experimental work has been expended on this problem, but relatively little theoretical effort. It is difficult because it spans an even wider range of conditions than the shock-initiation problem. Stewart [24] has considered the DDT problem for the case of small heat release. He considers a flame overtaken by a shock, and obtains a criterion for the minimum shock strength which will cause the flame to go over into a detonation.

Granularity. Modern solid explosives are granular materials, prepared by pressing powders. They can be initiated by much weaker shocks than those required for homogeneous explosives (liquids or single crystals). The bulk temperature produced by the initiating shock is much too low to start the reaction. So hot spots produced by void collapse or intergranular friction play a key role.

Analytic treatment of the granularity problem is in its infancy. Bdzil and Ferm [4] have recently done a careful perturbation treatment of a one-dimensional nonreactive model of granularity, in which the initial density is a periodic function of distance. One must get the phase relationships right for the multiple wave reflections from the density variations. To do this requires carrying the perturbation analysis to third order. The behavior of this system is quite complex. A number of interesting results appear, which we do not have space to describe here.

2. Analog.

2.1. Derivation. An analog for the nonreactive Euler equations has been known for a long time. It can be arrived at from the Euler equations by a series of assumptions and approximations; see, for example Whitham (1974, Chap. 2). We present here a slightly different "derivation" better suited to the present application.

Start with the nonreactive Euler equations, specialized from the reactive set (1.1)

(2.1a) $$\dot{\rho} + \rho u_x = 0,$$

(2.1b) $$\rho \dot{u} + p_x = 0,$$

(2.1c) $$\dot{e} + p\dot{v} = 0,$$

(2.1d) equation of state $p = p(\rho, e)$.

First neglect the dependence of the pressure on energy, so that the equation of state reduces to $p = p(\rho)$, and we no longer need the energy equation. Next eliminate the need for the momentum equation by assuming that u is some given function of ρ. The appropriate choice of the function $u(\rho)$ depends on the particular application. This leaves just the mass equation, which, with $u(\rho)$ a given function, is sufficient, by itself, to define $\rho(x,t)$. Next go over to material (Lagrangian) coordinates, replacing the spatial position x by a particle label h. A convenient choice for h is

$$dh = \rho(dx - u\,dt);$$

see, for example, Courant and Friedrichs [5, §18]. In these material coordinates (h,t) the mass equation reads

$$\rho_t + \rho^2 u_h = 0.$$

Finally, put this into conservation form, by choosing an appropriate function $Q(\rho)$ related to the function $u(\rho)$ that we used above to replace the momentum equation. We take for Q

(2.2) $$Q = Q(\rho) = \int \rho^2 u'(\rho)\,d\rho + C,$$

where C is a constant of integration. In terms of Q the mass equation reads

$$\rho_t + Q_h = 0.$$

An important conceptual step is to identify, for each quantity in the analog, the quantity in the physical system to which it most closely corresponds, and to choose the analog names and symbols accordingly. In making this correspondence we regard the analog as a given (constructed) mathematical object, and ignore the "derivation" by which it was obtained from the physical system. Thus, for example, in the "derivation" the quantity Q is the mass flux. But (as we shall see later) its role in the analog corresponds most closely to that of the pressure in the physical system. Therefore in the analog we call Q the pressure and give it the symbol p. Instead of $Q = Q(\rho)$ we write $p = p(\rho)$. This is the equation of state for the analog; it corresponds to the physical equation of state $p = p(\rho, e)$. We write the analog as

(2.3a) $$\rho_t + p_x = 0,$$

(2.3b) equation of state $p = p(\rho)$.

For density ρ and time t we retain the physical names and symbols. For the space coordinate we have reverted to the more familiar symbol x, even though in the analog it acts more like a particle label (and we speak of "particle x"). It will be important to keep this in mind, more so when we take up the reactive case.

The simplest choice for the equation of state which gives a nonlinear differential equation is

$$p(\rho) = \tfrac{1}{2}\rho^2,$$

which gives for the equation of motion the familiar

$$\rho_t + \rho\rho_x = 0,$$

the simplest nonlinear hyperbolic partial differential equation. A shock moving into an initial state (p_i, ρ_i) satisfies

$$D = \frac{p - p_i}{\rho - \rho_i},$$

where (p, ρ) is the state just behind the shock. To simplify things, we will always choose for our initial state $p_i = \rho_i = 0$, the origin of the p–ρ plane. (This corresponds roughly to the so-called strong-shock approximation in the physical system.) With this initial state the shock relation becomes

(2.4) $$D = p/\rho.$$

For fixed D, this gives the Rayleigh line in the p–ρ plane; its slope is proportional to D. Figure 2.1 shows the equation of state, which also serves as the shock Hugoniot (an instance of degeneracy in the analog), and two Rayleigh lines. This may be compared with Fig. 1.3 for the physical system. The topology is the same, but the abscissa which makes the shock relation (2.4) a straight line for the analog is ρ instead of v. It is this correspondence, among other things, which leads to the identification of Q as the mathematical analog of the pressure.

The characteristic form of (2.3) is

$$\rho = \text{constant} \quad \text{on} \quad dx/dt = p'(\rho).$$

We have only the $(+)$, or forward acoustic family of characteristics, propagating with sound speed $p'(\rho)$. The $(-)$ family of characteristics of the

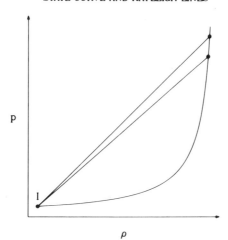

FIG. 2.1. *State curve and Rayleigh lines for the analog.*

physical system is absent. This simplifies shock problems considerably: signals overtaking the shock do not reflect from it, because the $(-)$ characteristics are absent. The shock does not act as a floating boundary which affects the flow behind it, but is simply fitted in after this flow field has been found.

The next step is to extend the nonreactive analog to include chemical reaction. To the original single dependent variable ρ we add a second dependent variable, the degree of reaction λ. We let the equation of state depend on both dependent variables so that it reads

$$p = p(\rho, \lambda).$$

We now have two dependent variables, but only one PDE. To complete the system we need to add a rate equation. In material coordinates (h, t) the physical rate equation (1.1d) is

$$\lambda_t = r.$$

We take this over as is (we chose to use material instead of spatial coordinates above so we could do this), and let the rate r, like the pressure p, depend on both ρ and λ.

Collecting our results, we have for our analog with reaction

(2.5a) $\qquad\qquad\qquad \rho_t + p_x = 0,$

(2.5b) $\qquad\qquad\qquad \lambda_t = r,$

(2.5c) $\qquad\qquad$ equation of state $p = p(\rho, \lambda),$

(2.5d) $\qquad\qquad$ reaction rate $r = r(\rho, \lambda),$

(2.5e) $\qquad\qquad\qquad$ shock $D = p/\rho.$

The number of dependent variables has been reduced from four (u, ρ, e, λ) to two (ρ, λ). The three original equations of motion (for mass, motion, and energy) have been replaced by the single conservation law (2.5a). The rate equation (in material coordinates) of the original is taken over unchanged. The constitutive relations still consist of an equation of state and a reaction rate, but each has two arguments instead of three. The system (2.5) is hyperbolic with two families of characteristics: the $(+)$, or forward acoustic, family, and the (0), or particle-path, family. The $(-)$, or backward acoustic, family of the original is absent. Initial and boundary conditions will be discussed in § 4.

2.2. Mechanical model. A simple mechanical model may be of help in thinking about the analog. The model is simply a line of cars travelling single file along a road, as shown in Fig. 2.2, Let the density ρ be the number of cars per unit length, and $u(x)$ the velocity of car x. In the continuum ap-

MECHANICAL MODEL

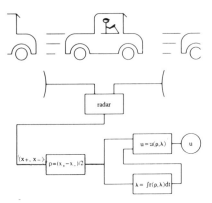

FIG. 2.2. *Mechanical model for the reactive analog: cars on a highway.*

proximation, the density of cars is governed by the physical mass-conservation equation (1.1a).

To avoid worrying about momentum and energy, we suppose that the mass of each car is negligible in comparison with the horsepower of its engine, so that its driver can set his velocity u instantaneously to any desired value, according to some prescribed recipe. We equip each car with a radar set, which can measure instantaneously the distance to its immmediate neighbors ahead and behind, and thus obtain an approximation to the local density ρ.

To obtain the nonreactive analog, we furnish each driver with some functional prescription $u(\rho)$ (the same for all) by which he is to set his velocity u at all times from the current value of the density ρ available from the radar.

To add reaction, we install in each car an integrator which integrates the differential equation

$$\lambda_t = r(\rho, \lambda),$$

with a prescribed rate function r. The variable λ is the state variable of the integrator itself, and the current value of ρ is fed into the integrator from the car's radar. The driver's velocity prescription now includes λ, the state of the integrator, as well as ρ from the radar, and takes the form of a prescribed function $u = u(\rho, \lambda)$. This gives us the reactive analog.

2.3. Equation of state and rate. Perhaps the simplest choice for the equation of state, which we take as a standard, is

(2.6) $$p(\rho, \lambda) = \tfrac{1}{2}(\rho^2 + q\lambda),$$

obtained by adding the term $\tfrac{1}{2}q\lambda$ to the commonly used $p = \tfrac{1}{2}\rho^2$ of the nonreactive analog. The constant q represents the heat of reaction. The function (2.6) is a one-parameter (λ) family of curves in the p–ρ plane, as shown in Fig. 2.3. There is a p–ρ curve for each value of λ. The differential of the pressure is

(2.7)
$$dp = c\,d\rho + \sigma\,d\lambda,$$
$$\text{frozen sound speed } c = p_\rho = \rho,$$
$$\text{thermicity } \sigma = p_\lambda = \tfrac{1}{2}q.$$

The second equality indicates the particular values for our standard equation of state (2.6). The *frozen sound speed* c is the (+) characteristic speed. The *thermicity* σ measures the heat of reaction, or more precisely in the general case, the effect of reaction on pressure. The system is exothermic if σ is positive and endothermic if σ is negative. For positive q, the state curves are displaced upwards and to the left as λ increases, as shown. These state curves play the part of the so-called partial-reaction Hugoniots of the physical system.

The shock relation (2.4) is the same for the reactive case. Throughout we make the tacit stipulation that λ does not change through the shock (because the shock jump is instantaneous while the reaction rate is finite). We take $\lambda_i = 0$ throughout.

The simplest choice for the rate is a function of λ alone. A common choice

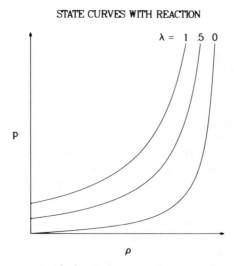

FIG. 2.3. *State curves with chemical reaction for positive heat of reaction q.*

is

(2.8) $$r = 2(1 - \lambda)^{1/2},$$

which has the convenient property that the reaction is complete in finite (unit) time. Although there is no temperature in the analog, we could perhaps simulate an Arrhenius factor for some purposes with a term like

$$e^{\rho^*/\rho},$$

with "activation density" ρ^*.

2.4. Variations. We can incorporate more complicated chemistry by adding additional λ's, one for each reaction, plus another rate equation like (2.5b) for each. The argument λ in the rate would then be replaced by λ_1, $\lambda_2, \cdots, \lambda_n$, and the term $q\lambda$ in the equation of state by $\sum q_i \lambda_i$. By choosing a suitable rate function we can also have two-way reactions like

$$A \rightleftarrows B$$

with an equilibrium composition. The analog has nothing like a thermodynamics with a free energy whose minimization gives the equilibrium composition, but we can choose a rate function $r(\rho, \lambda) = 0$, patterned after that of the physical system, whose solution gives an equilibrium composition $\lambda_e(\rho)$.

To do this we define an equilibrium constant

$$K(\rho) = se^{-h/\rho},$$

where

$$h = -\text{sign}(q) \, |q|^{1/2}.$$

The multiplier s here is like the entropy factor $e^{\Delta S/R}$ of the physical system, and the exponential $e^{-h/\rho}$ corresponds to the thermal factor $e^{-\Delta H/RT}$. The equilibrium composition λ_e is the solution of

$$\pi(\lambda) \equiv \frac{\lambda}{1 - \lambda} = K(\rho).$$

Using a mass action form for the rate

$$r = k_f(\rho)x_A - k_b(\rho)x_B,$$

with forward and backward rate constants k_f and k_b multiplying the concentrations x_A and x_B of reactant and product, and applying the equilibrium condition that at equilibrium we have $K = k_f/k_b$, we find for the rate

$$r(\rho, \lambda) = k_f(1 - \lambda)\left[1 - \frac{\pi(\lambda)}{\pi(\lambda_e)}\right],$$

where $\pi(\lambda_e) = K(\rho)$. This is essentially the physical result, with T replaced by ρ.

A detonation wave can be powered by the density change of a phase transition as well as by the heat release of a chemical reaction, see Rabie, Fowles, and Fickett [23]. We can obtain a good representation of such systems by using our standard equation of state (2.6) (with q now related to the difference in density between the two phases), and an appropriate choice for the equilibrium composition relation $\lambda_e(\rho)$.

We can represent the effects of transport properties by taking the Burgers form for the equation of motion

$$\rho_t + p_x = \nu\rho_{xx},$$

obtained from (2.5a) by adding the viscous or diffusive term $\nu\rho_{xx}$ on the right. This turns out to have all the desired properties for treating detonations or phase transitions in a viscous medium.

The analog is not applicable to flows in more than one dimension in any general way, but certain symmetric flows can be mocked up by adding a suitable ad hoc term to the right side of the equation of motion (2.5a). Thus for example, the effects of slight lateral expansion behind a detonation wave in a tube with heavy compressible walls are nicely represented by the addition of a constant term on the right. It appears that adding the term ρ/x on the right instead of a constant should give a useful representation of spherically symmetric flow.

2.5. Limitations. The analog has three main limitations: the absence of the $(-)$ family of characteristics, the limitation to one-dimensional flow, and the absence of thermal effects.

In the absence of the $(-)$ family of characteristics, waves propagate in one direction only. Problems requiring an explicit account of reflected waves cannot be treated. But there are many problems in which the only well defined waves are forward-going ones, modified, to be sure, by information traveling back along the $(-)$ characteristics from some extended reflection process. Here the analog gives a good account of the main features. The classic example is a rarefaction wave overtaking a shock. Another example is the problem of initiation of detonation by a shock described in §4.

The absence of thermal effects—energy and entropy—in the analog is one of the major factors in its simplicity, and is usually not a serious defect. The analog has only one state curve (at fixed composition); we are not troubled by the distinction between isotherms, isentropes, and Hugoniots. For strong enough shocks, of course, the irreversible heating becomes an important effect, which the analog omits. If such effects are to be studied, the analog will not do. But it does give a good qualitative rendering of the other important features in many problems where the presence of thermal effects complicates the problem considerably in the physical system.

Finally, although the analog is not applicable to flows in more than one dimension in any general way, certain symmetric flows can be mocked up by adding a suitable ad hoc term to the right side of the equation of motion (2.5a) as noted above.

3. Applications. In this section we give an overview of some successful applications of the analog. *All of the results presented are for the analog*, but they faithfully mirror those of the physical system.

3.1. Frozen and equilibrium sound speeds. Consider a system with a two-way reaction $A \rightleftarrows B$. Such a system exhibits two different sound speeds— the so-called frozen and equilibrium speeds. The frozen sound speed is the high-frequency limit in which the vibration is so fast that there is no time for the chemical composition to change. The equilibrium sound speed is the low-frequency limit in which the composition shifts to keep up with the state displacement so that the equilibrium composition is maintained at all times. Straightforward analysis of the governing equations shows that the frozen sound speed is the characteristic speed. The significance of the equilibrium sound speed is revealed by wave-operator analysis, as described, for example, in Whitham [27, Chap. 10]. It turns out to be a subcharacteristic velocity which is always less than the characteristic velocity. It is more like a group velocity—the velocity of propagation of the bulk of the energy in a disturbance.

3.2. Waves into an initial equilibrium state. Consider the behavior of a rarefaction or compression wave moving into the same material with two-way reaction $A \rightleftarrows B$, initially resting quietly in a state of chemical equilibrium.

The rarefaction is the simplest case. At early times the flow is described by the frozen equation of state, and at late times by the equilibrium equation of state. The transition from one to the other takes some tens of reaction times. Initially the wave head moves at the frozen sound speed. As time goes on the leading part of the wave continues to propagate with frozen sound speed, but its amplitude decays exponentially with time, becoming an exponentially decaying precursor to the effective wave head which propagates at the slower equilibrium sound speed.

For the compression case, consider a shock driven by a constant rear-boundary density ρ_b (analogous to a constant-velocity piston in the physical system), Fig. 3.1. Initially the shock is flat-topped, the reaction not having had time to proceed to appreciable extent. As time proceeds, a reaction zone

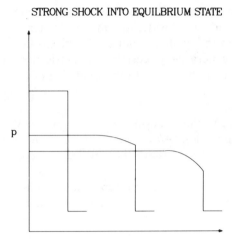

FIG. 3.1. *Strong shock running into a chemical equilibrium state; successive snapshots.*

forms behind the shock; in this zone the composition is shifting to its new equilibrium value. Ultimately we have a steady state with constant shock velocity and unchanging reaction-zone profile. In this problem the equilibrium sound speed becomes important only if ρ_b is below a certain critical value. In this case we have the situation shown in Fig. 3.2. The initial discontinuous shock jump decays to zero strength, and the final steady wave is a "fully dispersed" or "diffuse" shock propagating at a velocity a little above the equilibrium sound speed. The profile resembles that of a shock in a viscous medium, but here the "smearing" of the shock is caused by chemistry rather than viscosity.

3.3. Simplest steady detonation. In Fig. 1.4, we displayed the structure of a steady detonation wave in the physical system.[2] To review: at the head of the wave is a constant-velocity shock. The jump in state through the shock (high temperature in the physical system) turns on the reaction, resulting in a reaction zone attached to the shock. The reaction zone is steady in a frame moving with the shock. Attached to the end of the reaction zone (where reaction is complete) is some following flow, which may be nonsteady. Typically this following flow is a rarefaction wave which matches the high pressure at the end of the reaction zone to the much lower pressure at the rear boundary. We will be concerned here only with the shock and reaction zone.

---

[2] The reader is reminded that all results of this section (3) are for the analog. The conceptual framework for the steady detonation here is patterned after the standard ZND model of the physical system.

WEAK SHOCK INTO EQUILBRIUM STATE

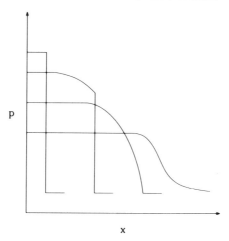

FIG. 3.2. *Weak shock running into a chemical equilibrium state.*

The steady solution for the analog is conveniently diagrammed in the p–ρ plane, Fig. 3.3 (compare Fig. 1.5 for the physical system). The state curves for no reaction ($\lambda = 0$) and for complete reaction ($\lambda = 1$) are shown. Two Rayleigh lines are also shown, the lower one for the unsupported, or Chapman–Jouguet, detonation, and the upper one for an overdriven detonation. (These two cases will be discussed in more detail in §4.) In either

DETONATION STATES

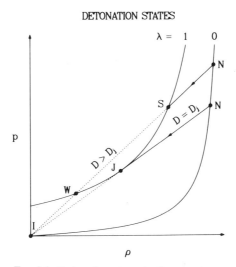

FIG. 3.3. *Detonation states in the p–ρ plane.*

case, the shock is a jump from the initial state I to a point N on the no-reaction curve. As a particle passes through the reaction zone, its state point slides down the Rayleigh line, covering the portion shown as a solid line, and ending up at the complete-reaction point J or S.

The unsupported or CJ detonation, which is not "pushed" from behind, is the most common case. It is defined by the condition that its Rayleigh line be tangent to the complete-reaction state curve. Thus both its propagation velocity (slope of the Rayleigh line) and the state at complete reaction depend only on the equation of state of the completely reacted material, and not on that of the partially reacted material, or on the reaction rate. The flow at point J is sonic so that the head of a following rarefaction wave can coincide with the end of the reaction zone, as shown in Fig. 1.4.

3.4. More complicated steady detonations. The analog readily accommodates the addition of other properties to the fluid model, such as additional chemical reactions, viscosity, or phase change. The study of such models reveals the possibility of a different type of steady solution, called the eigenvalue detonation. A necessary condition for the eigenvalue type of detonation is that the system contain some endothermic or dissipative process which, roughly speaking, cancels out the heat release from the main exothermic reaction at some point. In the p–ρ diagram, some partial-reaction state curves (not shown in Fig. 3.3) lie above the complete-reaction curve. The Rayleigh line for the unsupported eigenvalue detonation is like the upper one in Fig. 3.3; it intersects the complete-reaction curve in two points S and W. In a manner which requires more explanation than we have space for here, the state point arrives at the lower intersection W (for weak point) at the end of the reaction zone. The flow is supersonic there, so the reaction zone runs away from disturbances trying to catch it from behind. The eigenvalue detonation runs faster than the normal one, but has a lower final pressure. The propagation velocity of the eigenvalue detonation is determined by an eigenvalue solution of the ordinary differential equations governing the steady flow—hence the name. In sharp contrast to the normal case, the velocity and final state depend on the details of the reaction rate and of the equation of state of the mixture of reactants and products for all values of λ.

Steady detonation solutions have been obtained in the analog for the following systems:
1. One two-way reaction.
2. Two one-way exothermic reactions.
3. Two one-way reactions, one exothermic and one endothermic.
4. Two exothermic reactions, one one-way and one two-way.
5. One one-way exothermic reaction with viscosity.
6. Phase transition.

7. Phase transition with viscosity.
8. Detonation in a tube with slight lateral expansion.

In cases 6 and 7, the detonation is powered by the density change of a phase transition instead of by a chemical reaction. In case 8, the effects of the lateral expansion are put into the problem by adding an ad hoc constant source term, as noted earlier. Cases 3 through 7 have an eigenvalue solution for at least some choices of the reaction rate. In case 4 the required endothermicity enters when the change in state induced by the first reaction drives the second past its equilibrium point, so that it becomes endothermic as it returns to equilibrium. In case 8, the solution is always of the eigenvalue type, with the lateral expansion serving as the effectively endothermic process.

The analog's solutions for all of these problems are qualitatively the same as those for the physical system, and exhibit the expected eigenvalue solutions.

4. Basics. In this section we set out the basic concepts we will need in the presentations of the next two sections: the stability of the steady detonation in §5, and the initiation of detonation by a shock wave in §6.

4.1. Scaling. The governing equations (2.5), with our standard equation of state, admit a simple scaling. For a constant boundary condition, the values chosen for the rate multiplier and the heat of reaction q have no other effect than to set the time- and distance scales. We choose the rate multiplier to give unit reaction time in the steady solution of interest. The CJ detonation velocity is $D_j^2 = q$. We can choose $q = 1$ giving $D_j = 1$, or some other value of q which gives a convenient value of D for the steady solution of interest.

4.2. The piston problem. Recall the piston problem in the physical system, Fig. 1.1. The particle $x = 0$ in the analog plays the same role as the piston in the physical system. We call this the *boundary*. On the boundary we specify $\rho(t)$ and call this the *boundary condition* $\rho_b(t)$. The flow is determined by the given initial and boundary conditions

initial condition $\rho_i(x)$ on $t = 0$,

boundary condition $\rho_b(t)$ on $x = 0$.

For $\rho_i(x)$ we will always take $\rho_i = 0$, which gives the strong-shock jump relation (2.4).

If $\rho_b(t)$ is a positive step function we have, in the nonreactive case, a flattopped shock like that of Fig. 1.2. Figure 4.1 shows the t–x plane, with the shock path and a $(+)$ characteristic. The $(+)$ characteristics have positive slope; they convey information from the boundary up to the shock. If

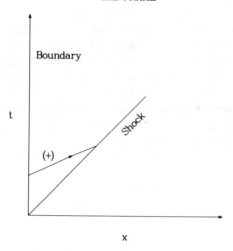

Fig. 4.1. *Shock and (+) characteristic in the laboratory frame.*

we add chemical reaction to the system, we will still have the leading shock, but its velocity will change with time, and the (+) characteristics will be curved.

4.3. Shock frame. It will be convenient to work in what we shall call the *shock frame*, an accelerated frame in which time is measured from the shock. In the transformation to this frame x is unchanged, and t is replaced by a new time variable τ. The transformation is

$$\tau(t, x) = t - t_s(x),$$

where $t_s(x)$ is the time of arrival of the shock at particle x, that is

$$t_s(x) = \int_0^x \frac{dx'}{D(x')}.$$

For constant shock velocity D, this is just

$$t_s(x) = x/D.$$

Partial derivatives transform as follows:

$$f_t \to f_\tau,$$
$$f_x \to f_x - D^{-1} f_\tau.$$

The governing differential equations (2.5a, b) become in the new frame

(4.1a) $$(D\rho - p)_\tau + Dp_x = 0,$$

(4.1b) $$\lambda_\tau = r.$$

Figure 4.2 is the shock-frame τ–x diagram corresponding to Fig. 4.1. The shock is now the horizontal axis. The boundary condition is still prescribed on $x = 0$ as before. The (0) characteristics (vertical lines) are unchanged. The (+) characteristics now have negative slope (as they must to reach the shock from the boundary).

4.4. Steady detonation. We seek the steady detonation solution, that is the solution in which the reaction zone (as in Fig. 1.4) is steady. By definition, in the steady solution the shock and its attached reaction zone propagate with constant velocity, and the history of every particle passing through the reaction zone is identical. We specify the steady propagation velocity D and seek the corresponding steady reaction-zone solution, thus obtaining a one-parameter (D) family of steady solutions. A specified boundary density ρ_b selects a particular member of this family, playing the same role as u_b in the physical system (see § 1).

We begin with the governing equations (4.1) in the shock frame. In the steady solution, D is a constant, and all derivatives with respect to x vanish (because all particle histories are identical). The equation of motion (4.1a)

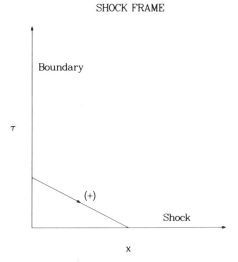

FIG. 4.2. *The shock frame. In this frame, time for each particle is measured from its passage through the shock, so that the shock becomes the horizontal axis.*

then becomes an exact differential

$$(D\rho - p)_\tau = 0,$$

giving

$$D\rho - p = \text{constant}$$

through the reaction zone. Now from the shock relation (2.4), the state just behind the shock lies on the Rayleigh line

$$p = D\rho;$$

thus the constant above is zero, and the entire reaction zone satisfies $p = D\rho$ and lies on this same Rayleigh line. Substituting the equation of state (2.6) for p gives us a quadratic equation relating ρ and λ

$$\rho^2 - 2D\rho + 2q\lambda = 0,$$

or

(4.2) $$\rho(\lambda;D) = D + \sqrt{D^2 - q\lambda}.$$

(We will use here only the (+) branch of the solution). As in the physical system, the minimum aceptable D is that for which the radical vanishes, that is $D = \sqrt{q}$. For this D the Rayleigh line is tangent to the complete-reaction state curve (lower line in Fig. 3.3). This is the so-called Chapman–Jouguet (CJ), or unsupported detonation, whose profile is that of Fig. 1.4. All larger values of D give Rayleigh lines of greater slope, such as the upper line in Fig. 3.3, and are called overdriven detonations. In either case, as the state point passes through the reaction zone, it slides down the Rayleigh line from the no-reaction point N on the $\lambda = 0$ curve to the complete-reaction point J or S on the $\lambda = 1$ curve.

The specified boundary condition selects the particular steady solution and determines the following flow behind the reaction zone. The discussion is essentially the same as for the physical system in §1. Take constant values of ρ_b. Let ρ_j be the value of ρ at the end of the reaction zone in the unsupported detonation (the tangent point J on the $\lambda = 1$ curve). For $\rho_b \leq \rho_j$ we have the unsupported detonation. For $\rho_b < \rho_j$ the following flow is a rarefaction wave which connects the density ρ_j with the lower value specified at the boundary. For $\rho_b = \rho_j$ the following flow is a constant state. For $\rho_b > \rho_j$ we have an overdriven detonation. The particular one is that for which ρ at the end of the reaction zone is equal to the given ρ_b. Here the following flow is a constant state, like that shown in Fig. 1.6. The steady solution $\rho(\lambda)$ for a given value of D is given by (4.2) above. To get the time history we turn to the rate equation (2.5c), which, with $\rho(\lambda)$ known, becomes an ordinary differential equation for $\lambda(\tau)$ along any particle path:

(4.3) $$\frac{d\lambda}{d\tau} = r(\rho(\lambda), \lambda), \quad \lambda = 0 \text{ at } \tau = 0.$$

4.5 Pressure form. It is convenient for some purposes to write the governing equations with p and λ as independent variables. This form has the nice property of displaying the effect of reaction through an explicit source term in the equation of motion.

To effect this transformation, use the inverse of the equation of state

$$\rho = \rho(p, \lambda) = (2p - q\lambda)^{1/2}$$

and its differential

$$d\rho = \rho_p dp + \rho_\lambda d\lambda.$$

Putting this into the equation of motion (2.5a), and using the derivative identities

$$\frac{\rho_\lambda}{\rho_p} = -p_\lambda = -\sigma, \quad \frac{1}{\rho_p} = p_\rho = c,$$

where p_λ and p_ρ are the partials of $p = p(\rho, \lambda)$, see (2.7), we obtain

(4.4) $$p_t + cp_x = \sigma r, \quad \lambda_t = r,$$

where $c = \rho$ and the equation of state and rate are $\rho = \rho(p, \lambda)$ and $r = r(p, \lambda)$. The term σr on the right is the reaction source term.

5. Stability. In the physical system, the hydrodynamic stability of the simplest steady detonation has received extensive study. If the activation energy (of an assumed Arrhenius rate) is large enough, the steady solution is found to be unstable to small disturbances. In nature, this manifests itself as a time-dependent cellular structure at the front of the wave, with the cell boundaries consisting of transverse waves moving across the main front.

A particularly interesting special case is restriction to one-dimensional (longitudinal) perturbations. Again with large enough activation energy the steady solution is found to be unstable; the detonation "gallops": by virtue of the restriction to one-dimensional perturbations, the front remains plane and all state variables are functions only of the longitudinal space variable x and the time t, but there is a large oscillation of the wave amplitude with time.

A calculated shock pressure history for a simplified ideal gas system is shown in Fig. 5.1. The oscillation amplitude is large: the peak height is nearly twice the steady value. The general shape—narrow spikes and wide

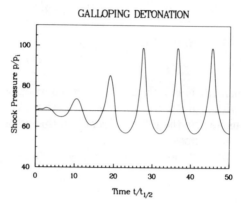

FIG. 5.1. *Shock-pressure history for a one-dimensional galloping detonation; $t_{1/2}$ is the half-reaction time in the steady solution and the horizontal line is the steady-solution shock pressure.*

troughs—is reminiscent of simple predator-prey (e.g. rabbit-coyote) models, see for example Fig. 4 of Hundhausen [18].

In nature, the one-dimensional gallop is rare; the three-dimensional mode (cellular front) appears to be preferred. However, in one particular situation the one-dimensional mode dominates: in the bow wave of a projectile fired into a detonable gas. The laser-schlieren photograph of Fig. 5.2 is an example. The (nearly) one-dimensional gallop takes place at the tip of the projectile and leaves its "tracks" behind in the form of vertical density striations. It may be that the one-dimensional mode is dominant here because any transverse waves which do form leak away to the sides.

The work of Alpert and Toong [1] gives an idea of the nature of the feedback mechanism of the longitudinal instability. A change at the shock propagates back to the reaction front along a particle path and a $(-)$ characteristic. These signals perturb the reaction front, and these perturbations propagate forward to the shock along $(+)$ characteristics. The backward path along the $(-)$ characteristic is less effective than that along the particle path. The results for the analog—in which the $(-)$ characteristics are absent—show that the particle path alone is sufficient to produce instability.

We present here an outline of the stability analysis for the analog. It should serve to present, in the simplest form, examples of the kinds of analysis appropriate to the study of detonation. It also provides insight into the nature of the instability.

We limit ourselves to the overdriven case. (In the unsupported case, the sonic point at the end of the reaction zone introduces complications like those occurring in transonic flow problems.) The steady solution is generated by a boundary density which reproduces the history of a particle in the steady

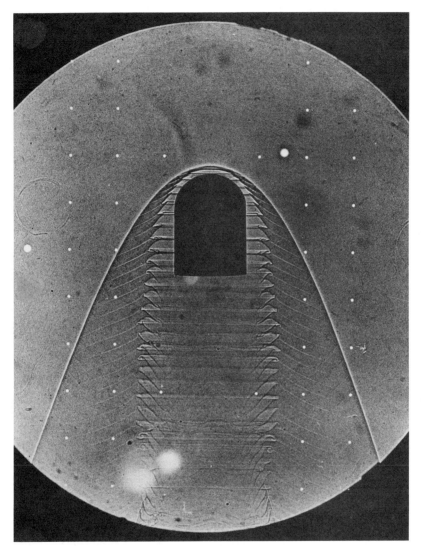

Fig. 5.2. *Galloping detonation at the tip of a projectile fired into a mixture of hydrogen and oxygen. From Lehr* [20].

solution. To study the stability, we add an arbitrary small perturbation to the boundary, and inquire whether the signals generated by it grow or decay.

We formulate the problem so as to resemble a standard two-port linear system, such as a passive R–C–L electrical network. We do this by selecting out of the flow field appropriate input and output functions which depend on a single variable. The applied boundary perturbation $x = 0$, a function of time alone, is the input function. To obtain an output function which depends on a single variable, we choose a particular path (curve in t–x space) through the flow field, and concentrate our attention on it. On such a path the perturbation is a function of a single variable; we take the perturbation along the path as the output function. The shock path is a convenient choice.

The standard linear-system form for the result is

(5.1)   (transform of output) = (transfer function) × (transform of input).

We are able to put our result in this form by applying a suitable transformation which changes the original input data on the boundary into effective input data along the shock path, so that both input and output lie on the same path. It is this effective input which appears in the standard form (5.1). The transforms are Laplace transforms, with x (distance along the shock) as the integration variable. The transformation is such that the effective input appears as the original input propagated from the boundary to the shock along the (+) characteristics, and damped by the chemical reaction.

Once we have obtained the standard form (5.1), we study the system in the usual way: examination of the transfer function yields the gain and phase shift of the system as a function of frequency, and determines its stability (the system is unstable if any poles in the right half-plane).

5.1. Reaction rate. The galloping-detonation calculation of Fig. 5.1 is for an ideal gas with a single first order reaction with Arrhenius temperature dependence. At this laboratory, our primary applications interest is in solid explosives. Instabilities of this type have been calculated, but not observed, in solids. By a happy coincidence, a rate function embodying one of the key features of reaction in granular solids (not applicable to gases) is also one of the simplest to treat theoretically. The remainder of our discussion here is for the analog with this rate function. The concept (Wackerle et al. [26]) is that passage through the shock impresses on each particle a "latent image" which influences its subsequent reaction rate. The idea is like that of the photographic process, in which the reaction rate of each silver halide grain during development depends on the light intensity it received earlier. Here, presumably, each granule of the explosive is damaged in some way by its passage through the shock, and this damage affects its subsequent reaction rate. To incorporate this idea into the analog in a simple way, we take the standard rate (2.8), and prefix it by a multiplier k, different for each particle, which depends on the shock strength experienced by the particle. We use the shock speed $D(x)$ as our measure of shock strength. The multiplier is then $k(x) = k[D(x)]$ and the rate becomes

(5.2) $r = r(x, \lambda) = k[D(x)][2(1 - \lambda)^{1/2}]$.

(In a real solid, there is also of course a local state dependence, which we omit here for simplicity.) We normalize this rate by choosing $k(D_0) = 1$, where D_0 is the propagation velocity of the steady solution. It turns out that the only property of the function $k(D)$ which affects the stability is its derivative at $D = D_0$. A possible form for $k(D)$ might be

$$k(D) = e^{\alpha(D - D_0)},$$

for which

$$k'(D_0) = \alpha.$$

Throughout, we work in the shock frame described in §4. We remark that $\lambda(\tau)$ and $r(\tau)$ can be found on any particle path in terms of the unknown multiplier $k(D)$, which is not known ahead of time because $D(x)$ is not known. We find these functions, which will be needed in the analysis, by solving the steady rate equation (4.1b) for $\lambda(\tau)$ and then evaluating $r(\tau) = r[\lambda(\tau)]$. The result is

(5.3a) $$\lambda(\tau) = 1 - (1 - \hat{\tau})^2,$$

(5.3b) $$r(\tau) = 2k(D)(1 - \hat{\tau}),$$

(5.3c) $$\hat{\tau} \equiv k(D)\tau.$$

5.2. Analysis. We outline the main steps in the analysis for the simplest case $k = $ constant in (5.2). For the more interesting but more complicated case of variable k, that is $k = k(D)$, we content ourselves with a brief description of the resulting changes.

For k constant, $\lambda(\tau)$ and $r(\tau)$ are known functions, the same for every particle path. This simplifies the analysis considerably. The governing equations in the shock frame are the pair (4.1) plus the constitutive relations

(5.4a) $$D\rho_\tau - p_\tau + Dp_x = 0,$$

(5.4b) $$\lambda_\tau = r,$$

(5.4c) equation of state $$p = \tfrac{1}{2}(\rho^2 + q\lambda),$$

(5.4d) reaction rate $$r = 2(1 - \lambda)^{1/2},$$

(5.4e) boundary $\rho_b(\tau)$ given.

The unknown shock velocity $D(x)$ appears in the coefficients of the equation of motion (5.4a). From the shock relation (2.5e) we have $D(x) = \tfrac{1}{2}\rho_s(x)$, where $\rho_s(x) = \rho(\tau = 0, x)$ is the shock density. Recall that in the analog, which lacks the $(-)$ characteristic family, the shock has no influence on the flow, and the shock relation is not needed as a floating boundary condition. The shock density $\rho_s(x)$ comes out as part of the solution and the shock relation gives $D(x)$ in terms of it. We will eventually replace $D(x)$ throughout by $\frac{1}{2}\rho_s(x)$, but will retain $D(x)$ for awhile to simplify the notation.

We begin by substituting the equation of state (5.4c) for the pressure p in the equation of motion (5.4a). The differential equations (5.4a, b) then read

(5.5a) $$(D - \rho)\rho_\tau + D\rho\rho_x = \tfrac{1}{2}q(\lambda_\tau - D\lambda_x),$$

(5.5b) $$\lambda_\tau = r.$$

Because r is a function of λ alone, we can solve the rate equation (5.5b) independently, reducing the system to a single equation. On the right side of (5.5a), λ_x is zero, and $\lambda_\tau = r(\tau)$ is the known function of τ given by (5.3b) with $k(D) = 1$. We have then the single governing equation

(5.6) $$(D - \rho)\rho_\tau + D\rho\rho_x = \tfrac{1}{2}qr(\tau),$$

where $r(\tau) = 2(1 - \tau)$ from (5.3b).

We linearize this equation, together with the shock relation and boundary condition, about the steady solution, writing our variables as

$$\rho(\tau, x) = \rho_0(\tau) + \epsilon\rho_1(\tau, x) + O(\epsilon^2) + \cdots,$$
$$D(x) = D_0 + \epsilon D_1(x) + O(\epsilon^2) + \cdots,$$
$$\rho_b(\tau) = \rho_{b0}(\tau) + \epsilon\rho_{b1}(\tau) + O(\epsilon^2) + \cdots,$$

etc., where sub 0 denotes the unperturbed flow. At $O(\epsilon)$ we find, after dividing by $(\rho_0 - D_0)$,

(5.7a) $$\rho_{1\tau} - v(\tau)\rho_{1x} = g(\tau)(\rho_1 - D_1),$$

(5.7b) $$\text{boundary} \quad \rho_{b1}(\tau) \quad \text{given},$$

where

(5.7c) $$v(\tau) = D_0\rho_0(\tau)/m(\tau),$$

(5.7d) $$g(\tau) = -\rho_0'(\tau)/m(\tau),$$

(5.7e) $$m(\tau) = \rho_0(\tau) - D_0 = \sqrt{D_0^2 - q\lambda_0(\tau)},$$

and the shock relation is

(5.7f) $$D_1(x) = \tfrac{1}{2}\rho_{s1}(x).$$

The coefficients $v(\tau)$ and $g(\tau)$ are known functions of τ, properties of the steady solution: $v(\tau)$ is the characteristic speed, and $g(\tau)$ is proportional to the (time-) gradient of ρ. The quantity $m(\tau)$ is related to the sonic character of the unperturbed flow (recall that $c = \rho$); note that the radical is the same one which appears in the steady solution (4.2).

The next step is to convert the PDE (5.7a) to an ODE by taking the Laplace transform on x. We use a tilde to denote the transform:

$$\tilde{f}(\tau, z) = L[f(\tau, x)] = \int_0^\infty e^{-zx} f(\tau, x)\, dx,$$

in particular,

$$\tilde{\rho}_1(\tau, z) = L[\rho_1(\tau, x)],$$

$$\tilde{D}_1(z) = L[D_1(x)].$$

The resulting ODE is

(5.8) $$\tilde{\rho}_{1\tau} = [zv(\tau) + g(\tau)]\tilde{\rho}_1 - v\rho_{b1}(\tau) - g(\tau)\tilde{D}_1.$$

The term with $\rho_{b1}(\tau)$ is the initial part of the transform of $\rho_1(\tau, x)$.

We now pause to explain our plan of attack from here on. As indicated in the section introduction, what we are after is the transfer function which relates input and output functions along the shock path. To find it, we solve (5.8) formally, with the unknown shock density $\tilde{\rho}_{s1} = 2\tilde{D}_1$ as the initial condition. Applying a boundedness condition to this solution places a constraint on the initial data, which gives us the desired equation for $\tilde{\rho}_{s1} = \tilde{\rho}_{s1}(z) = \tilde{\rho}_1(\tau = 0, z)$.

To obtain the formal solution, we integrate (5.8) from $\tau = 0$, temporarily treating the unknown initial value $\tilde{\rho}_{s1}(z)$ as known. The homogeneous part of (5.8) is

$$\tilde{\rho}_{1\tau} = [zv(\tau) + g(\tau)]\tilde{\rho}_1.$$

Its solution is

(5.9) $$\tilde{\rho}_1(\tau, z) = E(\tau, z),$$

where

$$E(\tau, z) = e^{zX(\tau) + G(\tau)},$$

$$X(\tau) = \int_0^\tau v(\tau')\, d\tau',$$

$$G(\tau) = \int_0^\tau g(\tau')\, d\tau'.$$

For later reference note that, since $v(\tau)$ is the characteristic velocity, $X(\tau)$ is distance along a characteristic. Using this homogeneous solution, we integrate from the shock to obtain the solution of the full equation (5.8) as

$$\tilde{\rho}_1(\tau, z) = E(\tau, z)\left[\frac{\tilde{\rho}_1(0, z)}{E(0, z)} - \int_0^\tau \frac{v(\tau')\rho_{b1}(\tau') + g(\tau')\tilde{D}_1(z)}{E(\tau', z)}\, d\tau'\right].$$

Note that $E(0, z) = 1$, and that the unknown initial value $\tilde{\rho}_1(0, z)$ is just $\tilde{\rho}_{s1}(z)$. We now replace $\tilde{D}_1(z)$ by $\frac{1}{2}\tilde{\rho}_{s1}(z)$ (from the transform of the shock relation (5.7f)), and write the solution as

(5.10) $$\tilde{\rho}_1(\tau, z) = E(\tau, z)\{\tilde{\rho}_{s1}(z)[-\tfrac{1}{2}J(\tau, z)] - J_i(\tau, z)\},$$

where

$$J(\tau, z) = \int_0^\tau \frac{g(\tau')}{E(\tau', z)}\, d\tau',$$

$$J_i(\tau, z) = \int_0^\tau \frac{v(\tau')\rho_{b1}(\tau')}{E(\tau', z)}\, d\tau'.$$

We now apply the boundedness condition to obtain an equation for $\tilde{\rho}_{s1}(\tau, z)$. For the transform to exist, the solution must be bounded as $\tau \to \infty$. Since $E(\tau)$ increases without bound in this limit, the factor in braces must vanish in the same limit. (It can be shown that this is a sufficient as well as necessary condition for the boundedness of $\tilde{\rho}_1(\tau, z)$.) Imposing this requirement gives the desired equation for $\tilde{\rho}_{s1}(z)$ as

$$\tilde{\rho}_{s1}(z) = \frac{J_i(\infty, z)}{1 - \tfrac{1}{2}J(\infty, z)}.$$

We now make a change of variable in the integrals which converts them to Laplace transforms in x and in effect transports the boundary data to the shock in the manner described in the section introduction. Examining the integrals in (5.10) and the definition of $E(\tau, z)$ in (5.9), we see that these integrals will become Laplace transforms if we replace τ' by X as the integration variable. We will of course need the function $\tau(X)$, the inverse of $X(\tau)$, to express the argument τ' of the integrand functions in terms of X.

Before writing down the result we discuss the physical significance of the transformation and make a slight change of notation. Consider an integral along the shock path with x as the integration variable. With each point x on the shock we associate a point $\hat{\tau}$ on the boundary through the inverse function $\tau(x)$, that is $\hat{\tau} = \tau(X = x)$. From the definition of X in (5.9), we see that the shock point x and its corresponding point $\hat{\tau}$ on the boundary are connected by a $(+)$ characteristic, as shown in Fig. 5.3. The function $X(\tau)$ is a transformation that projects an image of the boundary onto the shock by propagating elements along the $(+)$ characteristics. We call the integration variable x instead of X and denote by $\hat{\tau}(x)$ the inverse $\tau(X)$ of the function $X(\tau)$. With new names I and I_i, our integrals $J(\infty, z)$ and $J_i(\infty, z)$ become the Laplace transforms

(5.11)
$$I(z) = J(\infty, z) = \int_0^\infty e^{-zx} e^{-G[\hat{\tau}(x)]} \{g[\hat{\tau}(x)]/v[\hat{\tau}(x)]\}\, dx,$$

$$I_i(z) = J_i(\infty, z) = \int_0^\infty e^{-zx} e^{-G[\hat{\tau}(x)]} \rho_{b1}[\hat{\tau}(x)]\, dx.$$

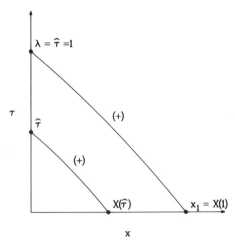

Fig. 5.3. *Imaging of the boundary on the shock (shock frame).*

Our solution for the shock density now has the desired form (5.1). In terms of the I's it is

$$(5.12) \qquad \tilde{\rho}_{s1}(z) = T(z)I_i(z), \qquad T(z) = \frac{1}{1 - \frac{1}{2}I(z)}.$$

Here $T(z)$ is the transfer function whose properties we want to study. Looking at $I_i(z)$, we see that the effective input signal on the shock path is as earlier described: it is the image on the shock of the input ρ_{b1} on the boundary, propagated down to the shock along the ($+$) characteristics, and damped by the gradient term e^{-G}. Note that the integral $I(z)$ in the denominator depends on the function g, which is proportional to the gradient of the steady solution. For a steady solution with zero gradient, as, for example, in the nonreactive case, $I(z)$ would vanish.

For the more interesting latent-image rate with nonconstant $k(D)$, the analysis is a little more involved. We cannot solve the rate equation independently at the outset, because of the dependence of its multiplier on D. But we can solve it for $\lambda[\tau, D(x)]$ along each particle path, as given by (5.3). The terms λ_τ and λ_x on the right side of (5.5) can be expressed in terms of this solution, yielding terms depending on $D(x)$ and its first derivative. These terms give rise to two additional terms on the right in the inhomogeneous part of the linearized PDE (5.7), and thus to two additional integrals in the transfer function, which becomes

$$1/T(z) = 1 - \tfrac{1}{2}\{I(z) + \tfrac{1}{2}q[I_a(z) - I_b(z)]\},$$

where

$$I_a(z) = L\{e^{-G}k'(D_0)[2(r_0 - 1)]/mv\},$$

$$I_b(z) = L[e^{-G}k'(D_0)D_0(r_0 - \tfrac{1}{2}r_0^2)/mv].$$

Here L denotes the Laplace transform as in (5.11), with x as the integration variable and $\hat{\tau}(x)$ as the argument of the functions G, m, v, and r_0 (recall that $r_0(\tau)$ is the known unperturbed rate function).

There are two parameters which affect the instability: the rate sensitivity $k'(D_0)$, and the degree of overdrive, that is the ratio D_0/D_j of the assumed steady-solution velocity to that of the unsupported wave. The results are relatively insensitive to the degree of overdrive. All of the results presented below are for $D_0/D_j = 2$.

5.3. Results. Stability is determined by examining the Nyquist diagram—the curve traced in the complex $1/T$ plane as z moves up the imaginary axis in the complex z plane. Calculation of each point of $T(z)$ requires evaluation of several quadratures. These were done by Runge–Kutta integration of the equivalent ODE's, with sixteen integration points for each period of the integrand.[3] For large $|z|$, an asymptotic expansion was used for $T(z)$. For presenting numerical results, we write z as

$$z/x_1 = \zeta + i\theta,$$

where the length scale x_1 is the length of the image of the boundary reaction zone on the shock, as indicated by the upper line in Fig. 5.3. The length of this image is about twice the length of the steady reaction zone in the laboratory frame.

We now turn to the results. We consider first the simpler case $k =$ constant. The Nyquist diagram is shown in the upper left corner of Fig. 5.4. The curve starts at $\theta = 0$ on the real axis, and the circles mark increasing values of θ at intervals of $\tfrac{1}{8}$ in $\theta/2\pi$. The steady solution is quite stable, and the transfer function differs little from unity. The maximum gain, which occurs at zero frequency, is only 1.07, and the maximum phase shift, near $\theta/2\pi = \tfrac{8}{3}$, is only two degrees.

The results for the case of variable k, that is $k = k(D)$ are more interesting. Nyquist diagrams for four values of $k' = k'(D_0)$ are shown in Fig. 5.4. For $k' = 0$, upper left, we have just the $k =$ constant case. As k' increases from zero, the curve expands, until it crosses the origin at $k' = 18.5$. This is the neutral-stability value; for larger k' the system is unstable. The first four

---

[3] Because both r_0 and g vanish outside the reaction zone, the integrals I, I_a, and I_b have a finite upper limit for a reaction complete in finite time.

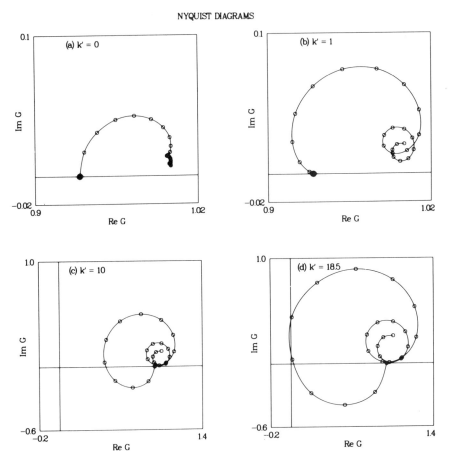

FIG. 5.4. Nyquist diagrams for different values of the rate-sensitivity parameter $k'(D_0)$, for $D_0/D_j = 2$.

poles of the transfer function at $k' = 18.5$ are

ζ	$\theta/2\pi$
0	0.49
−1.50	1.66
−1.97	2.69
−2.28	3.70

The spatial period of the fundamental (first pole above) is about $x_1/2$. The frequencies of the "harmonics" go up in steps which approach unity in the angular frequency $\theta/2\pi$. The higher the harmonic, the more rapidly it decays.

6. Initiation. We now turn to the problem of initiation by a plane shock. Typically the shock is produced by collision of a compressed-gas gun projectile with the explosive, Fig. 6.1. (The lateral dimension is large enough so that the flow in the central portion of the explosive remains one-dimensional in the time of interest.) The explosive is instrumented with pressure gauges, which have the form of thin plates. We idealize the gauges as fluid elements of the explosive, and the projectile as a constant-velocity piston.

We treat an analytically tractable limiting case. The limiting case is that of small heat release, realized by diluting the explosive with a suitable inert material while keeping the projectile velocity constant. The analysis supposes a dial on the heat of reaction and poses the question, "In the context of the shock-initiation problem, what are the first effects of adding exothermic chemistry to an inert flow?" The inert problem, which has a simple exact solution, is taken as the unperturbed case. We then consider a system with small heat of reaction, treating the heat of reaction as a small perturbation on the known inert flow. The small parameter ϵ for the analysis is proportional to the heat of reaction, being (roughly) the ratio of chemical to mechanical energy.

The reference or unperturbed case is the setup of Fig. 6.1 with chemical reaction formally included, but with the explosive made inert by setting its heat of reaction to zero. The reference flow is thus a constant-velocity flat-topped shock, with the pressure profile shown in Fig. 6.2. With the usual assumption that the shock turns on the reaction (the standard ZND model of detonation), a steady reaction zone follows the shock. The dashed curve of Fig. 6.2 shows a typical composition profile. Because the heat of reaction is zero, the reaction is decoupled from the flow, and has no effect on it. But the function describing the composition history of every particle (fluid element) passing through the steady reaction zone, which we call the *reference composition function,* is the key to our solution for nonzero heat of reaction.

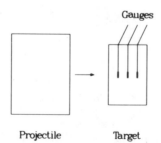

FIG. 6.1. *Experimental set up for study of the plane-shock initiation of detonation.*

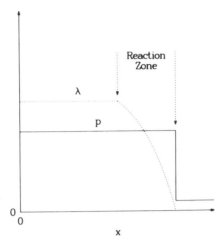

FIG. 6.2. *Reference flow.*

We sketch the analysis for the analog, and describe briefly the additional complications which arise in the physical system.

6.1. Analysis. We use subscript 0 to denote the shocked state of the reference flow, with D_0 the shock velocity. We define the small parameter ϵ as

(6.1) $$\epsilon = \tfrac{1}{2}q/D_0^2,$$

and choose a scaling which gives

$$\rho_0 = 2, \quad p_0 = 2, \quad D_0 = 1.$$

As in §5, we work in the shock frame, with its time variable τ, which is zero for each particle when it enters the shock. The reference composition function mentioned above is denoted by $\lambda_0(\tau)$; it is the composition history of a particle passing through the unperturbed steady reaction zone, the solution of the rate equation in this steady flow:

$$\frac{d\lambda}{d\tau} = r(\rho_0, \lambda),$$

$$\lambda = 0 \quad \text{at } \tau = 0.$$

We use the equations in the form (4.4) with independent variables p and

λ. Scaled as indicated above, and written in the shock frame, they are

$$p_\tau - a p_x = -\epsilon b r,$$

$$\lambda_\tau = r,$$

(6.2)
$$a \equiv \frac{cD}{c - D},$$

$$b \equiv \frac{D}{c - D},$$

boundary $p = 2 + \epsilon\lambda$.

Here c is the sound speed, $c = \rho = 2(p - \epsilon\lambda)^{1/2}$, and the boundary condition is the value of p implied by the constant-density condition $\rho = \rho_0 = 2$. (The impact is at $t = x = 0$; the boundary condition applies for $t \geq 0$.) The shock velocity is given by $D^2 = p_s/2$. The $(+)$ characteristic speed in this frame is $-a$. The flow is subsonic throughout, with the denominator $c - D$ always positive.

We linearize these equations for small ϵ, writing

(6.3)
$$p(\tau, x) = p_0 + \epsilon p_1(\tau, x) + O(\epsilon^2) + \cdots,$$

etc. At $O(\epsilon)$, we have for the pressure

(6.4)
$$p_{1\tau} - a_0 p_{1x} = -b_0 r_0(\tau),$$

boundary $p_{1b}(\tau) = \lambda_0(\tau)$,

where $a_0 = 2$, $b_0 = 2$ are the reference values of a and b, and $r_0(\tau) = r[\rho_0, \lambda_0(\tau)]$ is the rate history in the reference flow.

In characteristic form, (6.4) is

(6.5) $\quad P(\tau, x) \equiv p_1(\tau, x) + \lambda_0(\tau) = C_1 \quad \text{on } x + 2\tau = C_2,$

with path constants C_1 and C_2 to be determined from the boundary data. The paths are the $(+)$ characteristics of the reference flow. For the nonreactive case, the path invariant P is just p_1, which is thus constant along the characteristic; here we have the additional term $\lambda_0(\tau)$, and we begin to see the role played by the reference composition history.

We find the solution by evaluating the constants C_1, and C_2 for each characteristic on the boundary $x = 0$. After transforming back to the laboratory frame, the result is

$$p_1(\tau, x) = 2\lambda_0(\tau_+) - \lambda_0(\tau_s),$$

(6.6)
$$\tau_s \equiv t - x,$$

$$\tau_+ \equiv t - x/2.$$

6.2. Discussion. The pressure is $p = p_0 + \epsilon p_1$, so the pressure perturbation is proportional to ϵ, the scaled heat of reaction. The solution (6.6) is a superposition of two traveling waves. The amplitude function for both is the reference composition history $\lambda_0(\tau)$. The system may be thought of as an amplifier or converter of gain ϵ, which converts the composition signal of the reference flow into two pressure signals (travelling waves).

The two travelling waves are the steady and transient parts of the solution. Let us call them the $(+)$ wave and the s wave. The $(+)$ wave (argument τ_+) is the transient. It travels at the $(+)$ characteristic speed, which is twice the shock speed, and disappears by overtaking the shock. If we take a reaction which is complete in unit time, then we have the picture of Fig. 6.3: the $(+)$ characteristic through the complete-reaction point on the boundary $x = 0$ forms the upper boundary of the transient part of the flow.

The s wave (argument τ_s) travels at shock speed (recall that $D_0 = 1$), and thus constitutes the steady part of the flow. In the unshaded part of Fig. 6.3 we have the steady solution

(6.7) $$p_1 = 2 - \lambda_0(\tau_s),$$

the first term $\lambda_0(\tau_+)$ having become unity there.

The straightforward application of regular perturbation theory to hyperbolic problems ordinarily exhibits two far-field defects: wave-steepening effects are left out, and wave-position errors increase with time. The wave-steepening effects here are unimportant in the transient phase because it is over so quickly, and in the steady phase because it contains no interior

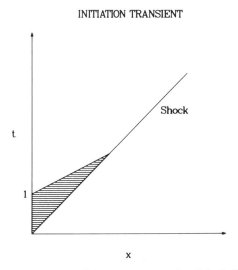

FIG. 6.3. *Extent of the initiation transient (shaded).*

shocks (the steady solution can be calculated exactly given the rate function). Wave-position errors do affect things, but in a fairly obvious way. We can use our solution (6.6) for p to adjust the shock and characteristic paths for a corrected value of the sound speed. When this is done, we find that the shape of the gauge histories is changed very little. The main effect is on the gauge timing: The shock runs faster and encounters each gauge sooner. The increase in shock speed is proportional to ϵ.

Our first order solution is independent of the sensitivity of the reaction rate to the state, that is, to the dependence of r on ρ. (In the physical system the state dependence is typically a dependence on temperature T through an Arrhenius term.) A preliminary look suggests that the analysis would have to be extended to third order to see an interesting dependence of the solution on the state-dependence of the rate.

6.3. Physical system. In the physical system the $(-)$, or back-facing, characteristics are also present; this complicates the problem considerably. A forward-going acoustic signal overtaking the shock is reflected from it. Both the sign and the magnitude of the reflected wave depend on the material. For the polytropic-gas equation of state

(6.8) $$p = (\gamma - 1)\rho(e + q\lambda),$$

for example, the reflection coefficient is a monotone function of γ. It is positive (a compression reflects as a compression) for $\gamma > 2$, negative for $\gamma < 2$, and zero for $\gamma = 2$. In effect, the analysis must take into account multiple reflections between the boundary and the shock, as shown in Fig. 6.4. This gives rise to a recursive equation for the boundary pressure, which can be converted to an infinite series.

There are invariants similar to (6.5) along both $(+)$ and $(-)$ characteristics.[4] The (suitably scaled) solution for p_1 (in the laboratory frame) is

(6.9)
$$2p_1(t, x) = -(a_+ - a_-)\lambda_0(\tau_s)$$
$$+ p_{1b}(\tau_+) + a_+\lambda_0(\tau_+)$$
$$+ p_{1b}(\tau_-) - a_-\lambda_0(\tau_-),$$
$$\tau_+ \equiv t - x/\rho_0 c_0,$$
$$\tau_- \equiv t + x/\rho_0 c_0,$$
$$\tau_s \equiv t - x.$$

The constants a_+ and a_- are proportional to the slopes of the $(+)$ and $(-)$ characteristics of the reference flow in the shock frame. The function p_{1b} is

---

[4] These take the form of the usual Riemann invariants plus a λ_0 term.

MULTIPLE REFLECTIONS

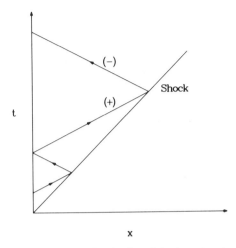

FIG. 6.4. *Multiple reflections between the shock and the boundary in the physical system.*

the piston pressure, given by

(6.10)
$$p_{1b}(y) = a_-\lambda_0(y) + (a_+ - a_-) \sum_{n=1}^{\infty} A^n \lambda_0(a^n y),$$
$$a \equiv a_-/a_+,$$

where A is the reflection coefficient for an acoustic signal reflecting off the shock. The constant a is less than one. The arguments $a^n y$ of λ_0 in the series represent the successive reflections of Fig. 6.4.

Here we have three travelling waves. The steady wave (argument τ_s) and the (+) wave (argument τ_+) play the same role as the corresponding ones in the analog. The (−) wave, (argument τ_-) travelling along the (−) characteristics, represents the reflection of the (+) wave from the shock. Although the piston pressure takes a formally infinite time to reach its final value, it is in practice found to be quite close after two or three reaction times. The travelling waves are similar to those of the analog, but have tails which reflect the delayed approach of the boundary pressure to its steady value.

6.4. Results. To illustrate the physical solution (6.9), we use the polytropic-gas equation of state (6.8), and a reaction rate which mocks up a branching-chain mechanism with an induction zone. Calculated pressure-gauge histories are shown in Fig. 6.5 for three values of γ.

For $\gamma = 1.2$, representing a gas, the reflection coefficient A is negative,

INITIATION RESULTS

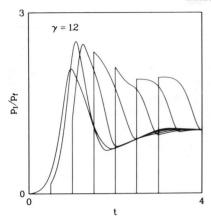

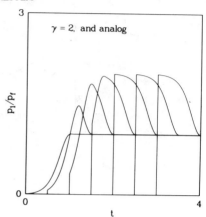

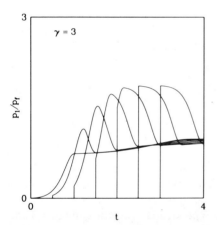

FIG. 6.5. *Pressure-gauge histories for three values of* γ *in the physical system. The ordinate is the pressure perturbation function* p_1, *normalized by its final (complete-reaction) steady-solution value* p_f. *For* $\gamma = 2$, *the physical result is identical to that for the analog.*

and the terms in the series in the expression (6.10) for the boundary pressure alternate in sign, giving extremes in the boundary-gauge history.

For $\gamma = 2$, an intermediate value which might be thought of as representative of liquid explosives, the reflection coefficient A vanishes, and the $(-)$ wave in (6.9) is not present. We find, in fact, that for this value of γ the physical solution (6.9) *becomes identical* to the analog solution (6.6). The boundary pressure reaches its final value at $t = 1$, when reaction is complete, and the transient phase is complete in finite time, as indicated in Fig. 6.3.

For $\gamma = 3$, representative of a solid explosive, the reflection coefficient is positive, and of smaller magnitude than for $\gamma = 1.2$, and the boundary pressure history is monotone. Note that for both $\gamma = 1.2$ and $\gamma = 3$, the steady state is closely approached in about four reaction times.

Notes. An excellent semi-popular overview of explosives and detonation for the technical audience is Davis [7]. Aside from this, there are three main background references: For nonreactive flow, Whitham [27, Part I, particularly Chaps. 1–4]; for reactive-flow theory and its specialization to detonation, and for the phenomenology of steady detonation, Fickett and Davis [14]; for a simplified presentation of reactive flow and detonation theory with the mathematical analog as a vehicle, Fickett [13]. A description of the analog and a brief introduction to classical detonation theory is given in Fickett [11]. For the rest of this section, we abbreviate Fickett and Davis [14] as FD, and Fickett [13] as F.

1. *Physical system.* The Euler equations are given in FD, §4A and Appendix 4C. The shock relations are described in Appendix 4C of this same source. A discussion of the probable importance of the transport terms of the Navier–Stokes equations in condensed explosives may be found in Eyring et al. [10].

An extensive treatment of detonation theory is given in FD. The elements of the theory and the ZND model are discussed in Chapter 2. An extensive catalog of a variety of steady solutions, including a number of eigenvalue detonations, are given in Chapter 5.

The stability of the steady solution is discussed extensively in FD, Chapters 6 and 7. For an application of geometrical acoustics in a reacting flow field, see FD, §6B2. For related material on sound waves in a reacting medium, see Toong [25, §10.3–6]. For the effects of granularity in a solid explosive, see Davis [7], Fickett [12, Refs. 1–5], and Johansson and Persson [19, §3.1].

2. *Analog.* The initial archival publication of the analog was in Fickett [11]. Essentially the same idea was proposed independently by Majda [21]. For a discussion of the Lagrangian vs. Eulerian routes for constructing it, see F, Appendix B, and for the chemical "thermodynamics" of the analog, see F, §3B.

3. *Applications.* For the basics of frozen vs. equilibrium sound speed, see F, §3E, and FD, §4A. For waves moving into an equilibrium state, see F, Chap. 4, and FD, §4C.

The simplest steady detonation is described in F, §5A and 5B, and FD, §5A. Several of the more complicated steady detonations in the physical system are described in FD, Chap. 5. For detonation powered by a phase transition, see Rabie, Fowles and Fickett [23]. For the steady detonation with lateral divergence, see FD, §5G and Bdzil [3]. For the analog, both the phase transition and lateral divergence are treated in F, Chap. 6.

4. *Basics*. Even though the analog is in a sense just a mathematical construct, it is conceptually convenient to assign a consistent set of units to the various quantities. This topic, and the scaling of the equations, is discussed in F, §3B.

We have found the shock frame, in which the time origin for each particle is the instant of its passage through the shock, the most convenient for our work. It is not new, but is less commonly used than the alternative of taking the position of the shock as the spatial origin. For an example of the latter, see Erpenbeck [9] and FD, §6A.

For more extended discussion of steady solutions of the equations of motion, see FD, §4A5, and F, §3H and Chap. 5.

5. *Stability*. The most careful and extensive study of the hydrodynamic stability of the steady solution is the work of Erpenbeck in the 1960's. The entire subject is reviewed in FD, Chap. 6, and also in Toong [25, §10-3]. The treatment of the one-dimensional "galloping" instability in the analog is given in F, Chap. 7. The analysis there is presented in steps of increasing complexity, and should be a good introduction for anyone wishing to study the work on the physical system. The reader who would like some background in the analysis of linear systems may consult standard texts, such as Frederick and Carlson [16]. See also F, Appendices D and E, where the elements of linear-systems theory are reviewed and applied to the harmonic oscillator as an example.

A feature of the stability analysis which has caused some puzzlement in the past (and may be responsible for some errors) may be worth mentioning. In simple situations not involving chemical reaction, one can use the steady frame of the *unperturbed* shock for the analysis, rather than the accelerated frame of the *perturbed* shock. An example is the problem described in Whitham [27, §8.1]. A flat-topped shock propagating through a constant (cross-sectional) area channel encounters a section of slowly varying area; the change in shock velocity due to this perturbation is to be calculated (in the quasi-one-dimensional approximation). Here the steady frame attached to the unperturbed, constant-velocity shock can be used, and this considerably simplifies the analysis. In the detonation problem, use of this frame results in the omission of important terms in the transfer function, and some of the work on the physical system appears to be in error for this reason. The difference between the detonation and channel problems is that in the detonation problem, the steady flow about which the perturbation is made—the steady reaction zone—has a gradient. If a gradient is present, it is necessary to place the shock at its correct position (to the order of the analysis), and this requires the use of the accelerated frame attached to the perturbed shock.

6. *Initiation*. The analysis of the initiation problem is given in detail in Fickett [12].

REFERENCES

[1] R. L. ALPERT AND T. Y. TOONG, *Periodicity in exothermic hypersonic flows about blunt projectiles*, Astronautica Acta, 17 (1972), pp. 539–560.
[2] J. B. BDZIL, *Perturbation methods applied to problems in detonation physics*, in Sixth Symposium on Detonation, Office of Naval Research, Arlington, VA, 1976, pp. 352–370.
[3] ———, *Steady-state two dimensional detonation*, J. Fluid Mech., 108 (1981), pp. 195–226.
[4] J. D. BDZIL AND E. FERM, *Shock properties in an Euler fluid with a periodic initial density*, 1984, unpublished.
[5] R. COURANT AND K. O. FRIEDRICHS, *Supersonic-Flow and Shock Waves*, Interscience, New York, 1948.
[6] M. COWPERTHWAITE, *Model solutions for the shock initiation of condensed explosives*, in Symposium on High Dynamic Pressure. Commisariat à l'Energie Atomique, Paris, 1978, pp. 201–211.
[7] W. C. DAVIS, *High explosives: the interaction of chemistry and mechanics*, Los Alamos Science, 2 (1981), pp. 48–75.
[8] R. ENGELKE AND J. B. BDZIL, *A study of the steady-state reaction structure of a homogeneous and a heterogeneous explosive*, Phys. Fluids, 26 (1983), pp. 1210–1221.
[9] J. J. ERPENBECK, *Stability of steady-state equilibrium detonations*, Phys. Fluids, 5 (1962), pp. 604–614.
[10] H. EYRING, R. E. POWELL, G. H. DUFFY AND R. R. PARLIN, *The stability of detonation*, Chem. Rev., 45 (1949), pp. 69–181.
[11] W. FICKETT, *Detonation in miniature*, Amer. J. Phys., 47 (1979), pp. 1050–1059.
[12] ———, *Shock initiation of detonation in a dilute explosive*, Phys. Fluids, 27 (1984), pp. 94–105.
[13] ———, *Detonation in Miniature*, Univ. California Press, Berkeley, 1985.
[14] W. FICKETT AND W. C. DAVIS, *Detonation*, Univ. California Press, Berkeley, 1979.
[15] G. R. FOWLES, *Stimulated and spontaneous emission of acoustic waves from shock fronts*, Phys. Fluids, 24 (1981), pp. 220–227.
[16] D. K. FREDERICK AND A. B. CARLSON, *Linear Systems in Communication and Control*, John Wiley, New York, 1971.
[17] D. D. HOLM AND J. D. LOGAN, *Self-similar detonation waves*, J. Phys. A. Math. General, 16 (1983), pp. 2035–2047.
[18] J. R. HUNDHAUSEN, *Advanced ordinary differential equations in an engineering environment: a computer assisted approach*, Amer. Math. Monthly, 87 (1980), pp. 662–669.
[19] C. H. JOHANSSON AND P. A. PERSSON, *Detonics of High Explosives*, Academic Press, London, 1970.
[20] H. F. LEHR, *Experiments on shock-induced combustion*, Astronautica Acta, 17 (1972), pp. 589–597.
[21] A. MAJDA, *A qualitative model for dynamic combustion*, SIAM J. Appl. Math., 41 (1981), pp. 70–93.
[22] A. MAJDA AND R. R. ROSALES, *A theory for Mach stem formation in reacting shock fronts, I: The basic perturbation analysis*, SIAM J. Appl. Math., 43 (1983), pp. 1310–1334.
[23] R. L. RABIE, G. R. FOWLES AND W. FICKETT, *The polymorphic detonation*, Phys. Fluids, 22 (1979), pp. 422–435.
[24] D. S. STEWART, *Transition to detonation in a model problem*, J. Mécanique, 1985, in press.
[25] T. Y. TOONG, *Combustion Dynamics*, McGraw-Hill, New York, 1983.
[26] J. D. WACKERLE, R. L. RABIE, M. J. GINSBERG AND A. B. ANDERSON, *A shock initiation study of PBX-9404*, in Symposium on High Dynamic Pressures, Commisariat à l'Energie Atomique, Paris, 1978, pp. 127–138.
[27] G. B. WHITHAM, *Linear and Nonlinear Waves*, John Wiley, New York, 1974.

CHAPTER V

Finite Amplitude Waves in Combustible Gases

J. F. CLARKE

1. Introduction. The generation of pressure waves from the rapid liberation of chemical energy within a gaseous system is one of the most immediately noticeable and practically significant of the consequences of many combustion processes. Apart from anything else the velocity of the signal that carries the message that an energy releasing reaction has taken place somewhere within the system will be that of the local speed of sound or greater if, as is likely, the signal is a shock wave. One usually thinks of these events on a rather large scale, of a laboratory or a building or even perhaps whole building complexes, so that the damaging blast waves have long ago left behind the regions in which actual chemical reactions occur and are travelling through the chemically inert surrounding atmosphere.

Important as the study of phenomena of this type obviously is, it must be recognized that pressure waves are born *within* domains that sustain the combustion reaction, where they will interact in a direct fashion with the reactions themselves and a mutual dependence will exist. This interdependence of combustion reactions and gas dynamics is the theme of the present article. The matter is one of great significance for *any* transient combustion process and will have applications even in pseudo-steady phenomena such as detonation waves, whose small-scale cellular structure consists of a regular pattern of transversely travelling waves and reaction fronts.

A potentially important field of application is to behaviour within the cylinders of internal combustion engines, within which a wave system is superimposed by movement of the piston on the generally spatially uniform background whose features change continuously with time.

To provide some unity in the present essay and at the same time concentrate on the theme of reaction/wave interactions, the geometry of the flow field is deliberately kept as simple as possible. Without implying any immediate connection with the practical problem of the preceding paragraph the discussion will be restricted to plane waves generated by events at the left-hand boundary of the atmosphere. This situation is very like the piston problems of classical gas dynamics but one must note two important distinctions from this well understood state of affairs. The first concerns the possibility for driving the wave field by simple deposition of heat through a *fixed* boundary. Although one may carry out this process in a chemically inert gas, when the driving mechanism is simply the displacement effects of the heated boundary layer of gas at the end wall, the augmentation of the displacement by chemical energy release, and subsequent sustainment by flame propagation, is clearly of great interest. Limitations of space preclude a discussion of the details of these processes, which are (to the writer's knowledge) not yet fully worked through anyway, but the foundations laid in the sections to follow should make it easier to complete the task.

The second distinction, referred to above, between present reactive and classical inert piston problems, is in the character of the ambient atmosphere. The matter is taken up in some detail in §3 and so will not be described here. Suffice it to say that the ambient atmosphere in the present study is essentially one which is undergoing an exothermic reaction, and is therefore changing its character with time.

It will be assumed throughout that the disturbance field is created at the left-hand boundary of the half-space filled with reacting gas, and that the initial conditions specify a stagnant, spatially uniform and homogeneous atmosphere. This idealisation and the consequent need to study boundary-value problems gives one a greater measure of access to the wave/reaction processes than is afforded by an initial-value formulation. Unless one is fortunate in one's choice of initial values, there will usually be a necessity to work through a sequence of "boundary layer"-like events immediately following the initial instant, which allow the system to settle into some natural mode of behaviour, and which may obscure this behaviour with some quite awkward analysis before it can be revealed. These remarks apply particularly to the chemical-composition field. There is also a temptation with initial-value problems to try to model behaviour in unbounded domains, and to rationalise the wide range of possibilities for initial-value data by choosing, for example, a harmonic wave train of uniform wavelength; it is then often difficult to reconcile such choices with real behaviour in finite or partly bounded systems.

Although one can obviously criticise the idealisations inherent in the present model's geometry, prolonged reflection has convinced the writer that its merits outweigh its demerits.

Before giving a resumé of the contents of this chapter and hopefully thereby making its motives and results clear, it is perhaps important to say explicitly what the work is *not* about. The interaction between flames and acoustical waves is a topic whose history goes back a long way and has many applications. The perennially instructive text by Markstein and coworkers offers ample evidence and would be very difficult to improve, even after twenty years. At the risk of trivialising such a crucial subject by attempting to summarise it briefly, one can think of these interactions as convective distortions or shakings (in the case of harmonic waves) of sheets of intense chemical activity namely the flames themselves. Changes in the rate and intensity of chemical activity within these sheets is brought about by their diffusive connections with their immediate surroundings through heat conduction and reactant and radical diffusion. There is no overt resultant generation of gas-dynamical waves by these changes in chemical activity, at least to leading order amongst the physical processes. Thus the present work, which does not consider diffusion *dominated* processes, supplements studies of the kind described by Markstein by making a mandatory, and reciprocal, relationship between gas-dynamical waves and chemical activity.

A start is made by writing down a set of conservation and related thermodynamic and chemical equations so as to provide an adequate summary of the main physical features of reactive compressible gas mixtures and, in particular, to exhibit the intrinsic capacity for wave propagation and its intimate connection with rates of chemical change. Section 3 discusses the ambient atmosphere and its properties, and derives equations that govern the propagation of *disturbances* to that atmosphere. Nondimensional quantities are defined in §4, which also gives some typical numerical values for such important dimensionless parameters as activation energy and Reynolds number. The finite reactive lifetime of the ambient atmosphere is emphasised through the medium of typical (associated) times-to-ignition of the atmosphere as a whole. Section 5 introduces the concept of waves that propagate in one direction only, as well as a set of semi-characteristic coordinates that make for a ready analysis of such processes. An important scale factor, that has a vital part to play in determining when and where general diffusive effects supervene, and where chemical activity is effectively suppressed, appears at this stage.

Section 6 introduces the idea of small perturbations, as well as the general solution technique that is based on limit processes and the construction of asymptotic parameter-perturbation series. The need for relationships (distinguished limits) between the several parameters that govern general behaviour is also seen here and several crucial choices for these distinguished limits are made. In particular disturbance-amplitude vis-à-vis activation-energy splits the problem into two halves labelled linear chemistry and nonlinear chemistry. In the first case the link between gas-dynamical and chemical activity, although crucial, is weaker than in the second. It is pointed out that behaviour in the ambient atmosphere may begin to exercise a dominant influence at times near to its (uniform) ignition.

The uni-directional waves are found (in §7) to obey an augmented version of the Burgers equation, the augmentation arising from coupling between waves and reactions *and* from changes in sound speed in the ambient atmosphere. For disturbances of truly (audible) acoustical amplitudes the

Burgers equation can be linearised and solutions show that behaviour near a wave head is quite complicated; of more practical significance is the identification of a wave frequency below which the plane waves are amplified by chemical activity to a greater extent than they are damped by viscous and heat-conduction effects. The essential dynamically "restless" and unstable character of a combustible atmosphere is revealed by this result.

With a rise in the amplitude of disturbances the nonlinear character of Burgers-like waves reasserts itself, and it is shown that viscous effects are predominantly confined to the interior of steep compression waves, or shocks. Furthermore the latter are found to propagate as essentially chemically-frozen Rankine–Hugoniot discontinuities, which grow in amplitude as a result of chemico-gas-dynamical coupling in the continuous wave fields that follow the shock front. With a sufficiently strong disturbance at early times it is possible for its amplitude to grow to the extent that nonlinear chemical response is provoked before the occurrence of ambient-atmosphere ignition. However two other, more notable, facts should be mentioned. First it should be remarked that all of the foregoing activity takes place with only very weak, second order, consumption of reactant species. Second, the onset of a more substantial consumption of reactant is the herald of a new type of wave behaviour. Uni-directional propagation is not sustained in these new circumstances and an unstable dispersive type of wave system is established. Section 8 goes into some detail in one case of this kind, when the breakdown of uni-directional propagation is inspired by the rapid evolution of events in the ambient atmosphere itself.

A further rise in the input-disturbance amplitude leads to nonlinear chemical response, but with reactant consumption still of second order. A most important new feature of these stronger disturbances is their capacity to advance and localise the process of ignition (christened "local explosion"). That is to say, ignition of the gas mixture is a process that can be propagated into the heart of the field at local sound (or even shock) speeds, with all the consequences that may follow from that fact. Once again an advance in the extent of reactant consumption that is associated with the local ignition process is itself associated with the generation of a new dispersive unstable wave field, with both directions of wave propagation now co-existing. There are several possible combinations of circumstances that can arise in the presence of local explosion and these are described and, to some extent, illustrated towards the end of §9, which is devoted to a study of the influences of nonlinear chemical activity.

The recurrence of equations that describe an unstable dispersive wave field is a feature of the whole subject as events proceed towards either local explosion or ambient-atmosphere ignition. The last parts of §9 go into some little detail in the very important case for which there is strong nonlinear involvement of the chemistry and the paper concludes, in §10, with the exploitation and solution of this equation to describe the onset of ignition behind a strong shock wave. It is shown that the wave field has the important task of fixing the induction time, in this case at the arithmetic mean of constant-volume and constant-pressure values.

In most real circumstances the ambient atmosphere will exhibit some degree of spatial nonuniformity in addition to its necessary changes with time. An idealisation that would help to resolve difficulties inherent in such a combination of ambient-atmosphere processes would be the assumption that spatial nonuniformity is the *only* feature of the ambient atmosphere (before the advent of perturbations). Such a system must involve convection and it is no simple task to work out the details of the relevant field which may, for example, be composed of a steady shock wave followed by a zone of reactive flow that quickly approaches a state of equilibrium, such as is found in the classical ZND model of a detonation wave. Perturbation of such a steady state field is tantamount to an enquiry into its stability, albeit perhaps to co-planar disturbances as a problem to begin with. The present approach can be adapted to such studies, and some early work of this kind is referred to in the References section at the end of the paper. Much remains to be done and the writer hopes that some avenues of approach will be revealed by the present essay.

Observations of real events have marched far ahead of explanatory, and particularly quantitative, theoretical modelling in many aspects of combustion and explosion phenomena, and this is particularly true of unsteady or transient processes. The need for some pre-conditioning of a combustible mixture, in the form of a gradient of temperature in the field, for example, so that the mixture shall be receptive to the processes that lead to deflagration-to-detonation transition, is well attested experimentally. The same is true of the necessity for *coherence* in the propagation of compression waves and the concomitant liberation of energy by combustion. For these reasons the present essay focusses on the primary physical processes of interaction or coupling between gas-dynamics and combustion reactions. It is axiomatic in the theory to be described below that a gas-dynamical wave of compression or expansion will provoke changes of reaction intensity within the gas through which it travels, and that these alterations of chemical activity make direct modifications to the wave strengths and so complete the "feedback" loop. Much understanding emerges from consideration of small amplitude waves. The essay proceeds by steps through wave/chemistry disturbance fields of larger and larger amplitudes and the important time, length and relative amplitude scales are identified at each stage. The *intrinsically* coherent character of wave field and combustion chemistry is apparent throughout, and the theory provides a base on which to build predictive structures (qualitative and quantitative) of transient gaseous combustion phenomena.

2. Basic equations and other generalities. It is important to have a clear idea of the basis and implications of the equations that will be used to illustrate the wave propagation problems that are the subject to this article. Accordingly this section is devoted to a brief but reasonably complete statement of the model and its interpretation.

2.1. Conservation and other equations. The equations of conservation of mass, momentum, energy and species for a one-dimensional unsteady flow are

(2.1) $$\frac{D\rho}{Dt} + \rho u_x = 0,$$

(2.2) $$\rho \frac{Du}{Dt} + p_x = \tau_{vx},$$

(2.3) $$\rho \frac{Dh}{Dt} - \frac{Dp}{Dt} = -\mathcal{Q}_x + \tau_v u_x,$$

(2.4) $$\rho \frac{Dq}{Dt} = -\rho \mathcal{R} - g_x,$$

where u, ρ, p, h and q are respectively the gas velocity, density, pressure, enthalpy and mass fraction of the combustible species. The viscous stresses are written as τ_v, and

(2.5) $$\tau_v = \tfrac{4}{3} \eta \, u_x$$

where η is the dynamic viscosity coefficient. The quantity written as $\mathcal{Q}$ is the single component of the energy flux vector, namely

(2.6) $$\mathcal{Q} = -\lambda \theta_x + g\hat{Q} \equiv \mathcal{Q}_\theta + g\hat{Q},$$

where $\mathcal{Q}_\theta$, the thermal part of the energy flux vector, is defined for convenience. The quantity g is the diffusive mass flux of the combustible species; it is sufficiently accurate here to use Fick's law, so that

(2.7) $$g = -\rho \mathcal{D} \, q_x$$

where $\mathcal{D}$ is the mass diffusion coefficient, and $\hat{Q}$ is the energy of combustion per unit mass of gas mixture.

The reaction rate term $\mathcal{R}$ will be assumed to have the Arrhenius form

(2.8) $$\mathcal{R} = nW\mathcal{P} \exp(-\rho E_A/p) q^n,$$

where n is the order of the simple (irreversible) chemical reaction, W is the molecular weight of the combustible species, E_A is the activation energy of the reaction and $\mathcal{P}$ is a pre-exponential factor that has the dimensions of (time)$^{-1}$; for simplicity $\mathcal{P}$ will be assumed constant in the present analysis. It must be noted that E_A is an energy per unit mass and that the particular form of the exponential in (2.8) implies the assumption that the gas mixture has a constant molecular weight W_m, so that

(2.9) $$p = \rho(R/W_m)\theta$$

where θ is the absolute temperature. It is only possible for W_m to be exactly constant if *all* species have the same molecular weight, in which case $W_m = W$. However if the gas mixture contains only a little fuel, so that q is a small fraction, it is reasonable to think of W_m as effectively constant and so it will not necessarily be the case that W_m and W are the same. The energy equation (2.3) must of course be augmented by some additional thermodynamic information; in the present case it will be assumed that enthalpy h is given by the following simple caloric equation of state,

(2.10) $$h = C_{pf}\theta + q\hat{Q},$$

where C_{pf} is the (assumed constant) frozen specific heat at constant pressure.

Evidently the chemistry of the mixture has been subjected to a number of idealizations. In view of the complexity of the whole problem that is being studied here such idealizations will only produce minor variations on the main theme of the links and interactions between the dynamics of the gas and the inherent chemistry of combustion; the latter certainly has its principal features incorporated in the model via (2.3), (2.4) and (2.8) in particular.

2.2. Recognition of intrinsic wavelike behaviour. The whole theme of the present article is the coupling of chemistry and wave processes in a combustible atmosphere. The chemistry is already evident in the equations of §2.1 through the presence of $\mathcal{R}$, and it is therefore important to expose the underlying wavelike character of the fields described by these equations in as general a fashion as possible.

Equations (2.9) and (2.10) show that

(2.11) $$\rho\frac{Dh}{Dt} = \frac{\gamma}{\gamma - 1}\rho\frac{D}{Dt}(p/\rho) + \rho\hat{Q}\frac{Dq}{Dt}$$

where γ, the ratio of frozen specific heats, is given by

(2.12) $$\gamma/(\gamma - 1) = C_{pf}W_m/R.$$

Substituting (2.11) into (2.3) and making use of (2.1), (2.6) gives

(2.13) $$\frac{Dp}{Dt} + \rho a_f^2 u_x = (\gamma - 1)\{\rho\hat{Q}\mathcal{R} - \mathcal{Q}_{\theta x} + \tau_v u_x\},$$

where

(2.14) $$a_f^2 = \gamma p/\rho$$

is the square of the local frozen sound speed, a_f. It is important to remark on the general significance of the concept of a local frozen sound speed and, indeed, to note the validity of an equation like (2.13) under much more general circumstances than its derivation here would suggest.

In physical terms equation (2.13) displays the "heating" effects of chemical energy release, heat conduction and viscous dissipation (the three terms on the right-hand side of (2.13), respectively) on the relationship between pressure and volume changes for a fluid particle (the terms on the left-hand side of (2.3)).

Combination of $(\pm a_f)$ times (2.2) with (2.13) now shows that

(2.15) $$(p_t + [u \pm a_f] p_x) \pm \rho a_f(u_t + [u \pm a_f] u_x)$$
$$= (\gamma - 1) \{\rho \hat{Q} \mathcal{R} - \mathcal{Q}_{\theta x} + \tau_v u_x\} \pm a_f \tau_{vx},$$

where either upper or lower signs are to be taken together. The derivative terms in parentheses in (2.15) reveal the existence of the underlying wavelike structure in any disturbance field. The waves have local propagation speeds equal to $\pm a_f + u$, but such local wavelike behaviour may be heavily suppressed or modified in regions where the "diffusive" *non*-wavelike phenomena of heat conduction and viscous action cannot be neglected. Even when $\mathcal{Q}_\theta$ and τ_v are negligible the existence of chemical energy release must have an important influence on the wavelet propagation, and vice versa.

This latter comment can be given additional force if one combines (2.13) and (2.1), (2.4) to give

(2.16) $$\frac{Dp}{Dt} - a_f^2 \frac{D\rho}{Dt} + (\gamma - 1) \rho \hat{Q} \frac{Dq}{Dt} = 0,$$

where the transport effects have been omitted at this stage. If q does not change, the relationship (2.16) simply states that the rate of change of p with ρ for any fluid particle is equal to a_f^2. It can now be seen that one cannot account for the influence of chemical change on the system by simply modifying the "compressibility" $\{\partial p/\partial \rho\}_{\text{particle}}$ to account for differences in the chemical composition; as (2.16) shows, there is an essential addition source term that forces one to pay due attention to the *rates* at which such compositional changes are brought about. Although it is implicit that our model of combustion phenomena will require $\hat{Q} > 0$, there is no reason why $\hat{Q}$ at this stage of the analysis should be positive. It is therefore opportune to comment that one must anticipate very different influences from chemistry when the reaction is endothermic or exothermic.

3. Ambient atmosphere. If fuel and oxidant gases are mixed together, they must be reacting although, to be sure, the reaction process may be extremely slow if the temperature of the mixture is low (e.g. as in normal atmospheric conditions). The inference is that either spatial or temporal nonuniformity, or both, must exist even before one sets out to disturb the mixture with the intention of examining the way in which such disturbances behave.

There is much to be said for choosing the ambient atmosphere to be spatially uniform, largely because its behaviour is then described by quite simple versions of the set of conservation equations given in §2. It is important to observe that the ambient atmosphere must indeed satisfy the conservation and other equations. The consequences of spatial uniformity will be seen to demand constant density in the atmosphere. Any attempt to impose constancy of *pressure* on the atmosphere will lead to problems, such as the need to have an ambient flow speed that grows linearly, and therefore unboundedly, with distance from an arbitrary origin.

Under conditions of spatial uniformity the atmosphere must be changing, as time proceeds, towards a state of chemical equilibrium. In the present case, for which the irreversible reaction rate is given by (2.8), this state must be represented by a zero value of the mass fraction q and a cessation of all chemical activity. For present purposes this idealization, which effectively forbids the treatment of the influence of an equilibrium atmosphere on disturbances propagating through it, is acceptable. Any spatial nonuniformity can be considered as a perturbation, not necessarily small of course, to the atmosphere, and its individual development can then be examined in company with that of the atmosphere itself. There is a clear possibility for the existence of two grossly different time scales for the evolution of the atmosphere ("long") and of the perturbation ("short"); indeed this is very frequently what does happen with flames in premixed atmospheres and the two-time formulation is a rational way in which to circumvent the cold boundary difficulty encountered in steady flow models of flame propagation.

It is easy to visualize the initiation of disturbances to the temporally evolving but spatially uniform atmosphere through the movement of a piston (the reacting gas counterpart of the classical "piston problem" of gas dynamics). Alternatively disturbances could be created by a large and rapid increase in the temperature of a fixed solid boundary, modelling a local ignition source. Indeed it will be seen in what follows that there is some gain in simplicity of the theoretical boundary-value problem over the initial-value formulation.

3.1. Equations for the ambient atmosphere. If the values of the dependent variables in the spatially uniform atmosphere are distinguished by the use of a subscript zero, it is clear from the appropriate special forms of (2.1),

(2.2) that

(3.1) $$u_0 = 0,$$
$$\rho_0 = \rho_i = \text{constant}.$$

Then (2.4) and (2.15) show that

(3.2) $$\frac{dq_0}{dt} = -\mathcal{R}_0 = -nW\mathcal{P}\exp(-\rho_i E_A/p_0)q_0^n,$$

(3.3) $$\frac{dp_0}{dt} = \rho_i(\gamma - 1)\hat{Q}\mathcal{R}_0 = -\rho_i(\gamma - 1)\hat{Q}\frac{dq_0}{dt},$$

since q_0 and p_0 depend upon t only. Integration of (3.3) shows that

(3.4) $$p_0(t) + \rho_i(\gamma - 1)\hat{Q}q_0(t) = p_{0i} + \rho_i(\gamma - 1)\hat{Q}q_{0i} \equiv p_{0m}$$

where $(\ )_i$ indicates an initial value (at $t = 0$) and p_{0m} is the maximum ambient pressure, achieved when $q_0 = 0$ and all ambient reaction has ceased. Combination of (3.2)–(3.4) shows that

(3.5) $$\frac{dp_0}{dt} = [\rho_i(\gamma - 1)\hat{Q}]^{1-n} nW\mathcal{P}\exp(-\rho_i E_A/p_0)[p_{0m} - p_0]^n$$

which can be solved to find $p_0(t)$, with the initial value $p_0(0) = p_{0i}$.

Equation (3.5) is of most immediate importance for the reason that it defines a time interval over which the changes in the value of p_0 itself, and hence, by virtue of (3.4), of q_0 too, are comparatively modest. This time, often called the ignition time for the homogeneous atmosphere, but perhaps better described as an induction time, since the truly rapid energy releasing event that causes a precipitous rise in $p_0(t)$ actually occurs somewhat later in the sequence of spatially homogeneous events, can be expressed as follows. Defining ϵ, the activation energy number, by

(3.6) $$\epsilon \equiv p_{0i}/\rho_i E_A,$$

writing p_0 as $p_{0i} + \epsilon p_0^{(1)}(t)$ and retaining only terms of leading order on each side of (3.5), this equation shows that $p_0^{(1)}$ is logarithmically infinite when t is equal to a time t_I, the induction time. It can easily be shown that

(3.7) $$t_I = \epsilon\, p_{0i}/\rho_i(\gamma - 1)\hat{Q}\mathcal{R}_i.$$

It is important to observe that t_I only has real *physical* significance if ϵ is small, since only then can significant increases in the exponential factor $\exp(-p_{0i}/\epsilon p_0)$ occur before significant reductions in q_0, or equivalently in $[p_{0m} - p_0]$. Since it is advantageous in various parts of the analytical work that follows to use not t_I but γt_I, it is noted here that

(3.8) $$\gamma t_I = \epsilon\, a_{f0i}^2/(\gamma - 1)\hat{Q}\mathcal{R}_i = \{a_{f0i}^2/(\gamma - 1)\hat{Q}\}\{\epsilon\, e^{1/\epsilon}/nW\mathcal{P}q_{0i}^n\}.$$

3.2. Disturbances to the ambient atmosphere. Subtraction of (3.2) and (3.3) from (2.4) and (2.15), and recognition of the fact that x-derivatives of ambient atmosphere variables are zero, shows that

$$\frac{D}{Dt}(q - q_0) = -(\mathcal{R} - \mathcal{R}_0) - \frac{1}{\rho} g_x, \tag{3.9}$$

$$\begin{aligned}((p - p_0)_t + [u \pm a_f](p - p_0)_x) &\pm \rho a_f(u_t + [u \pm a_f]u_x) \\ &= (\gamma - 1)\hat{Q}(\rho\mathcal{R} - \rho_i\mathcal{R}_0) - (\gamma - 1)\{\mathcal{Q}_{\theta x} - \tau_v u_x\} \pm a_f \tau_{vx}.\end{aligned} \tag{3.10$\pm$}$$

Thus it follows that there is a reasonably straightforward set of equations that describes the behaviour of *disturbances*, $(p - p_0)$, $(q - q_0)$, ρ (or $(\rho - \rho_i)$ of course) and u to the ambient atmosphere; for example one could use (2.1) with (3.9) and (3.10$\pm$), or either of (3.10+ or $-$) can be dropped in favour of (2.2). Coefficients of the various rates of change of disturbance quantities with x or t involve both ambient and disturbed values, of course.

4. Dimensionless equations.

4.1. Identification of parameters. It is important to work in terms of dimensionless quantities in this kind of analysis. At this stage it is not crucial to have exactly the proper dimensional base on which to work (this can be uncovered as the analysis proceeds) and so one can simply choose the set of dimensionless quantities, indicated by an over bar ($\bar{}$), as follows

$$\begin{aligned}&\text{(a)} \quad p - p_0 = \rho_i a_{f0i}^2 \bar{p}, \qquad p_0 = \rho_i a_{f0i}^2 \bar{p}_0, \\ &\text{(b)} \quad \rho - \rho_i = \rho_i \bar{\rho}, \\ &\text{(c)} \quad u = a_{f0i} \bar{u}, \\ &\text{(d)} \quad a_f = a_{f0i} \bar{a}_f, \qquad a_{f0} = a_{f0i} \bar{a}_{f0} = a_{f0i} \sqrt{\gamma \bar{p}_0}, \\ &\text{(e)} \quad q - q_0 = q_{0i} \bar{q}, \qquad q_0 = q_{0i} \bar{q}_0, \\ &\text{(f)} \quad \eta = \eta_i \bar{\eta}, \quad \lambda = \lambda_i \bar{\lambda}, \quad \mathcal{D} = \mathcal{D}_i \bar{\mathcal{D}}, \\ &\text{(g)} \quad x = \gamma t_I a_{f0i} \bar{x}, \qquad t = \gamma t_I \bar{t}.\end{aligned} \tag{4.1}$$

The result in (4.1d) follows from the definitions and (2.14).

Equations (2.1) and (2.2), respectively, become

$$\bar{p}_{\bar{t}} + \overline{u\rho}_{\bar{x}} + (1 + \bar{\rho})\bar{u}_{\bar{x}} = 0, \tag{4.2}$$

$$(1 + \bar{\rho})\{\bar{u}_{\bar{t}} + \overline{uu}_{\bar{x}}\} + \bar{p}_{\bar{x}} = \frac{1}{\text{Re}}(\overline{\eta u}_{\bar{x}})_{\bar{x}}, \tag{4.3}$$

while (3.10±) transform into

$$(\bar{p}_{\bar{t}} + [\bar{u} + \bar{a}_f]\bar{p}_{\bar{x}}) \pm (1 + \bar{p})\bar{a}_f(\bar{u}_{\bar{t}} + [\bar{u} \pm \bar{a}_f]\bar{u}_{\bar{x}})$$

$$= \epsilon \exp\left\{\frac{1}{\epsilon}\left(1 - \frac{1}{\gamma \bar{p}_0}\right)\right\} \bar{q}_0^n$$

(4.4±) $\quad\quad\quad \times \left\{(1 + \bar{p}) \exp\left[\frac{1}{\epsilon}\left(\frac{1}{\gamma \bar{p}_0} - \frac{1 + \bar{p}}{\gamma \bar{p}_0 + \gamma \bar{p}}\right)\right]\left(1 + \frac{\bar{q}}{q_0}\right)^n - 1\right\}$

$$+ \frac{1}{\Pr \operatorname{Re}} (\lambda \bar{\theta}_{\bar{x}})_{\bar{x}} + \frac{\gamma - 1}{\operatorname{Re}} (\overline{\eta u u_{\bar{x}}}) \pm \frac{1}{\operatorname{Re}} \bar{a}_f (\overline{\eta u_{\bar{x}}})_{\bar{x}},$$

where

(4.5a,b) $\quad\quad \theta = \theta_{0i}\bar{\theta} = (p_{0i} W_m / \rho_i R)\bar{\theta}, \quad\quad \bar{\theta} = \gamma(\bar{p}_0 + \bar{p})/(1 + \bar{p}).$

The natural appearance of the activation energy number ϵ should be noted, together with the appearance of parameters Re and Pr; Re is a Reynolds number, Pr a Prandtl number, defined as follows:

(4.6a,b) $\quad\quad\quad \operatorname{Re} = \gamma t_I a_{f0i}^2 \rho_i / \tfrac{4}{3} \eta_i, \quad\quad \Pr = \tfrac{4}{3} \eta_i C_{pf} / \lambda_i.$

Equation (3.9) can be written in dimensionless terms as

$$\bar{q}_{\bar{t}} + \bar{u}\bar{q}_{\bar{x}} = -(\epsilon/\bar{Q}) \exp\left\{\frac{1}{\epsilon}\left(1 - \frac{1}{\gamma \bar{p}_0}\right)\right\} \bar{q}_0^n$$

(4.7) $\quad\quad\quad \times \left\{-1 + \exp\left[\frac{1}{\gamma \epsilon}\left(\frac{\bar{p} - \bar{p}_0 \bar{p}}{\bar{p}_0(\bar{p}_0 + \bar{p})}\right)\right]\left(1 + \frac{\bar{q}}{q_0}\right)^n\right\}$

$$+ \frac{1}{\operatorname{Sc} \operatorname{Re}(1 + \bar{p})}((1 + \bar{p})\overline{\mathcal{D}} \bar{q}_{\bar{x}})_{\bar{x}}$$

and two new parameters, Sc and $\bar{Q}$, make their appearance; Sc is a Schmidt number and $\bar{Q}$ a combustion energy number that measures the available combustion energy $(\gamma - 1)q_{0i}\hat{Q}$ against the "thermal" energy a_{f0i}^2. The quantities are defined as follows:

(4.8a,b) $\quad\quad\quad \operatorname{Sc} = \tfrac{4}{3} \eta_i / \rho_i \mathcal{D}_i, \quad\quad \bar{Q} = (\gamma - 1)q_{0i}\hat{Q}/a_{f0i}^2.$

Finally, the dimensionless forms of the ambient atmosphere equations (3.4) and (3.5) are

(4.9) $\quad\quad\quad\quad\quad\quad\quad \bar{p}_0 + \bar{Q}\bar{q}_0 = \bar{p}_{0m},$

(4.10) $\quad\quad\quad\quad\quad \dfrac{d}{dt} \bar{p}_0 = \epsilon \exp\left\{\dfrac{1}{\epsilon}\left(1 - \dfrac{1}{\gamma \bar{p}_0}\right)\right\} \bar{q}_0^n.$

From (2.14) and (4.1a) the initial condition for (4.10) is given by

(4.11) $$\bar{p}_0(0) = 1/\gamma \equiv \bar{p}_{0i}.$$

4.2. Some illustrative numbers. Of the various parameters that appear in the equations in §4.1, ϵ and Re^{-1} *may* be assumed to have very small values in what follows, but γ, Pr, Sc and $\bar{Q}$ will always be treated as of order of magnitude unity.

When t is zero, $\bar{q}_0$ is unity and so (4.9), (4.11) together show

(4.12) $$\bar{Q} = \frac{1}{\gamma}\left(\frac{\bar{p}_{0m}}{\bar{p}_{0i}} - 1\right),$$

since $\gamma\bar{p}_{0i}$ is equal to one. Thus, from (3.8) and (4.8b), and choosing $n = 1$,

(4.13) $$\gamma t_I = \gamma \epsilon e^{1/\epsilon}/W\mathcal{P}\,((\bar{p}_{0m}/\bar{p}_{0i}) - 1).$$

The group of terms $\frac{4}{3}\,\eta_i/\rho_i\,a_{f0i}^2$ that appears in Re in (4.6a) is of the order of the mean molecular collision interval, which is roughly of the same order of size as the pre-exponential factor $\mathcal{P}$. Accordingly

(4.14) $$\mathrm{Re} \approx t_I\mathcal{P} = \epsilon e^{1/\epsilon}/W\,((\bar{p}_{0m}/\bar{p}_{0i}) - 1).$$

Some representative numbers for $\bar{p}_{0m}/\bar{p}_{0i}$ and W are 5 and 30, respectively, and Table 1 lists some related values for Re and t_I, the latter on the assumption that $\mathcal{P}$ is 10^9 Hz, for given activation energy numbers, ϵ.

TABLE 1
Reynolds number Re and induction time T_2 (in seconds) for various values of activation energy number ϵ.

$1/\epsilon$	15	20	25	30	35	40	45
Re	1.8E + 3	2.0E + 5	2.4E + 7	3.0E + 9	3.8E + 11	4.9E + 13	6.5E + 15
$\mathrm{Re}^{-1/2}$	2.4E − 2	2.2E − 3	2.0E − 4	1.8E − 5	1.6E − 6	1.4E − 7	1.2E − 8
t_I, sec	1.8E − 6	2.0E − 4	2.4E − 2	3.0E + 2	3.8E + 2	4.9E + 4	6.5E + 6
$-\epsilon \ln \mathrm{Re}^{-1/2}$	0.17	0.31	0.34	0.36	0.38	0.39	0.41

5. Uni-directional waves. The analysis of waves that travel in only one direction is both informative and simpler than the more general case. Accordingly this restricted class of problems will be considered first.

5.1. Piston problems and independent variables. Assume that the ambient atmosphere is disturbed by the movement of a piston, that lies at $\bar{x} = 0$ for all times $\bar{t} \leq 0$. At the initial instant the piston will begin to move in some prescribed manner. Define an amplitude parameter σ and a time scaling parameter σ_1. For reasons that will become apparent as the analysis proceeds it is then advantageous to describe the piston path as follows;

(5.1) $$\bar{x} = \sigma_1\sigma^2 D(T/\sigma), \qquad \sigma_1 T = \bar{t},$$

where $D(T/\sigma)$ is a function whose order of magnitude is about unity. The dimensionless piston speed is therefore

(5.2) $$\bar{u}_p = \frac{d\bar{x}}{dE} = \sigma\,\frac{d[D(T/\sigma)]}{d(T/\sigma)} \equiv \sigma D'(T/\sigma),$$

where D' is likewise a quantity of unit order of magnitude. Both σ and σ_1 are adjustable constant parameters; evidently, from (5.2), σ will measure the amplitude of the gas velocity input to the system while (5.1) makes it clear that σ_1 will have a controlling influence on temporal rates of change that are imposed on the system by piston movement. Since one is not especially interested in variations over time scales much longer than t_I, it is clear that σ_1 can be restricted to be in the order class unity or less.

Since we shall be interested in the propagation of waves from the piston face out into the exothermically reacting gas that lies in the regions of $\bar{x}$-positive, it is clear that a coordinate $\bar{t} - \bar{x}$ will have a significant role to play. In particular it is helpful to define ξ via

(5.3) $$\sigma_1\sigma\xi = \bar{t} - \bar{x}.$$

Equation (4.4+) shows that wavelets propagate in the direction $\bar{x}$-positive along the family of characteristics β = constant, where

$$\left(\frac{\partial \bar{x}}{\partial \bar{t}}\right)_\beta = \bar{u} + \bar{a}_f.$$

In view of (5.1) and (5.3) this is equivalent to

(5.4) $$\sigma\left(\frac{\partial \xi}{\partial T}\right)_\beta = 1 - \bar{u} - \bar{a}_f = -\sigma\,\beta_T/\beta_\xi,$$

and it is even more helpful to use β and T as a pair of independent coordinates with which to describe the waves. Transformation from the $\bar{x}$, $\bar{t}$ pair of variables is simply made. For completeness the derivative relations are given here,

(5.5a) $$(f_{\bar{t}})_{\bar{x}} = \frac{1}{\sigma_1}(f_T)_\beta + \frac{1}{\sigma_1\sigma}\beta_\xi[\bar{u} + \bar{a}_f](f_\beta)_T,$$

(5.5b) $$(f_{\bar{x}})_{\bar{t}} = -\frac{1}{\sigma_1\sigma}\beta_\xi\,(f_\beta)_T,$$

where f is any relevant dependent variable, and it must be observed that the derivative β_ξ makes an essential appearance (cf. 5.4). Note that

(5.6) $$\beta_\xi \equiv \left(\frac{\partial \beta}{\partial \xi}\right)_T, \quad \beta_T \equiv \left(\frac{\partial \beta}{\partial T}\right)_\xi$$

in (5.4) and, indeed, throughout the present article. It is also opportune to remark that

(5.7) $$\xi_T \equiv \left(\frac{\partial \xi}{\partial T}\right)_\beta, \quad \xi_\beta \equiv \left(\frac{\partial \xi}{\partial \beta}\right)_T$$

here and from now on.

5.2. Equations in semi-characteristic coordinates. For obvious reasons a pair of variables such as β, T introduced in §5.1, is called semi-characteristic. (4.2), (3), (4±) and (7) in terms of such variables are

(5.8) $$\sigma \xi_\beta \bar{\rho}_T + \bar{a}_f \bar{\rho}_\beta - (1 + \bar{\rho}) \bar{u}_\beta = 0,$$

(5.9) $$(1 + \bar{p})\{\sigma \xi_\beta \bar{u}_T + \bar{a}_f \bar{u}_\beta\} - p_\beta = \frac{1}{\sigma_1 \sigma \operatorname{Re}}\left(\frac{1}{\xi_\beta}\overline{\eta} u_\beta\right)_\beta,$$

$$p_T + (1 + \bar{p})\bar{a}_f \bar{u}_T$$

(5.10) $$= \sigma_1 \epsilon \exp\left\{\frac{1}{\epsilon}\left(1 - \frac{1}{\gamma \bar{p}_0}\right)\right\} \bar{q}_0^n$$

$$\times \left\{(1 + \bar{p}) \exp\left[\frac{1}{\gamma \epsilon}\left(\frac{\bar{p} - \bar{p}_0 \bar{p}}{\bar{p}_0(\bar{p}_0 + \bar{p})}\right)\right]\left(1 + \frac{\bar{q}}{q_0}\right)^n - 1\right\}$$

$$+ \frac{1}{\sigma_1 \sigma^2 \operatorname{Re}\xi_\beta}\left\{\frac{1}{\Pr}\left(\frac{\bar{\lambda}}{\xi_\beta}\theta_\beta\right)_\beta + \bar{a}_f\left(\frac{\bar{\eta}}{\xi_\beta}\bar{u}_\beta\right)_\beta - \sigma_1 \sigma(\gamma - 1)\overline{\eta u} u_\beta\right\};$$

(5.11) $$\sigma \bar{q}_T + \bar{a}_f \bar{q}_\beta \frac{1}{\xi_\beta}$$

$$= -\sigma_1 \sigma(\epsilon/\bar{Q})\exp\left\{\frac{1}{\epsilon}\left(1 - \frac{1}{\gamma \bar{p}_0}\right)\right\}\bar{q}_0^n$$

$$\times \left\{\exp\left[\frac{1}{\gamma \epsilon}\left(\frac{\bar{p} - \bar{p}_0 \bar{p}}{\bar{p}_0(\bar{p}_0 + \bar{p})}\right)\right]\left(1 + \frac{\bar{q}}{q_0}\right)^n - 1\right\}$$

$$+ \frac{1}{\sigma_1 \sigma \operatorname{Re}\operatorname{Sc}(1 + \bar{p})\xi_\beta}\left((1 + \bar{p})\frac{\overline{\mathcal{D}}}{\xi_\beta}\bar{q}_\beta\right)_\beta.$$

Apart from the desirability of making the derivative β_ξ equal to unity, at least in the regions near to the piston face, the wavelet or characteristic parameter β is arbitrary at this stage. It can best be chosen to have the value

T/σ at the point where the wavelet emanates from the piston path on a distance-time, or $\bar{x}, \tilde{t}$, picture, so that the boundary conditions (5.1) and (5.2) become

(5.12) $\quad \bar{u}(T/\sigma, T) = \sigma D'(T/\sigma), \quad \xi = (T/\sigma) - \sigma D(T/\sigma), \quad \beta = T/\sigma,$

where the condition on ξ results from the combination of (5.1), (5.3). It is important to remember that in the present situation ξ is essentially a dependent variable, a function of T and β, to set alongside $\bar{\rho}, \bar{p}, \bar{u}$ and $\bar{q}$.

5.3. Influence of the scale factor ξ_β. The quantity ξ_β defined in (5.7) is a scale factor that describes the local size of a spatial interval in $\bar{x}$, for a given $\tilde{t}$ or T, as one moves from one wavelet, and hence one β-value, to another.

The effect of a reduction in the magnitude of ξ_β is worth noting. In the momentum equation (5.9) reductions in the size of ξ_β cause the transport (in that case viscous) effects to become more significant. The latter is also true for (5.10), although this equation shows that reductions in ξ_β need be less pronounced for their influence to be felt than is the case with the momentum equation. In other words, reductions in ξ_β will first be significant in the energy balance equation (5.10).

Both (5.10) and (5.11) show a diminishing role for the reaction-rate terms as ξ_β decreases. This indicates in general terms the existence of locally chemically frozen or near-frozen flows that involve sufficiently rapid local rates of change from one wavelet to another.

5.4. More on the character of ξ_β. It is clearly possible to go some way with an evaluation of ξ directly from (5.4) and (5.12). Thus (5.4) gives

$$\sigma \xi = \int_{\sigma\beta}^{T} [1 - \bar{a}_f(\beta, \tilde{T}) - \bar{u}(\beta, \tilde{T})] \, \partial \tilde{T} + f(\beta),$$

where $f(\beta)$ is a function to be determined from (5.12). It easily follows that

(5.13) $\quad \sigma(\xi - \beta + \sigma D(\beta)) = \int_{\sigma\beta}^{T} [1 - \bar{a}_f(\beta, \tilde{T}) - \bar{u}(\beta, \tilde{T})] \, \partial \tilde{T}.$

Differentiating (5.13) and making use of (5.12) again, it can now be shown that

(5.14) $\quad \xi_\beta = \bar{a}_f(\beta, T = \sigma\beta) - \frac{1}{\sigma} \int_{\sigma\beta}^{T} [\bar{a}_{f\beta}(\beta, \tilde{T}) + \bar{u}_\beta(\beta, \tilde{T})] \, \partial \tilde{T}.$

These results will be useful in the analysis that follows.

In choosing the various scale or gauge factors, particularly those for time and the wave position coordinate ξ (or, equivalently, β), the implicit decision has been taken to emphasise the existence of waves propagating in a single direction, $\bar{x}$-positive in this case, when the parameter σ is taken to be a small number. On the assumption that differentiations with respect to T at fixed β, and to β at fixed T, are all order-one operations, it is clear from (5.8), (5.9), for example, that change from one wavelet to another is emphasised over and above temporal rates of change *along* the wavelet under these circumstances. Such waves are sometimes called "slowly varying", although that nomenclature will not be adopted here. There is certainly a very wide range of conditions under which uni-directional waves exist. The analysis that follows is predicated on the assumption that they do at least do so for early times from the start of the process; the question of how long uni-directional propagation continues as the dominant mode is one that the subsequent analysis is able to answer.

6. Small perturbations and uni-directional waves. The equations displayed in §5.2 are certainly quite complicated, and do not encourage the view that any useful exact analytical solutions exist; certainly none have been found so far. One must therefore investigate the possibilities for approximate analysis, based on the general notion of "small" disturbances to the ambient atmosphere.

It is necessary to remark that "small" does *not* mean "linear", as will be demonstrated below. To be sure the linear equations that govern the behaviour of disturbances of very small, acoustical, amplitudes have a part to play in any study of gas dynamical waves. But they do not exploit the full range of analytical possibilities, which certainly encompasses weak-shock wave formation and development, and which is therefore essentially nonlinear in character. Quantification of words like "small" and "weak" will be achieved, as the analysis progresses, with the aid of parameter-perturbation methods.

6.1. Some general results for small perturbations. It will now be assumed that σ is a small parameter and that, as suggested by (5.12), $\bar{u}$ is $O(\sigma)$. It will also be assumed, for the present at least and until it is shown otherwise, that $\partial/\partial T$ and $\partial/\partial \beta$ are $O(1)$ operations; in other words $\bar{u}_T$ and $\bar{u}_\beta$ are $O(\bar{u})$. Furthermore, Re will be assumed to be suitably large; what constitutes suitability will emerge shortly.

Equations (5.8) and (5.9) suggest that both $\bar{p}$ and $\bar{\rho}$ must be $O(\sigma)$ and, accordingly, the following asymptotic developments are proposed,

(6.1) $\qquad \bar{\psi} \sim \sigma \psi^{(1)}(\beta, T), \qquad \psi = u, p, \rho,$

together with the limiting process

(6.2) $\qquad\qquad \sigma \to 0, \quad \beta, T \text{ fixed.}$

It follows from (2.14) and (4.1) that

(6.3) $\qquad \bar{a}_f \sim \bar{a}_{f0} + \tfrac{1}{2} \sigma \sqrt{\gamma} \, (\bar{p}_0^{-1/2} p^{(1)} - \bar{\rho}_0^{1/2} \rho^{(1)}),$

whence, provided that $\sigma_1 \text{Re} \to \infty$ (NB Table 1) and $\xi_\beta > 0$ as $\sigma \to 0$, (5.8) and (5.9) give

(6.4) $$p^{(1)} = \bar{a}_{f0}^2 \rho^{(1)} = \bar{a}_{f0} u^{(1)},$$

provided that the arbitrary functions of time that appear as a result of the integrations with respect to β can be set equal to zero. Since there must be no disturbances for at least some values of β (in other words, the disturbances can only have existed for a finite time interval), it is evident that this is indeed so.

6.2. Distinguished limits. A group of terms that appears in the index of the exponentials in (5.10) and (5.11) is (NB the results in §6.1)

(6.5) $$\frac{1}{\gamma \epsilon} \frac{\bar{p} - \bar{p}_0 \bar{p}}{\bar{p}_0(\bar{p}_0 + \bar{p})} \sim \frac{\sigma (\gamma - 1)}{\epsilon} \frac{u^{(1)}}{\bar{a}_{f0}^3},$$

and it is evidently necessary to make some decision about the quotient σ/ϵ before taking the limit $\sigma \to 0$. The relative sizes of σ and ϵ will distinguish different domains of physical and mathematical behaviour, hence the descriptive phrase "distinguished limits", and there are evidently a number of such distinctions to be drawn in the present multi-parameter problem (although note the opening sentence in §4.2).

In general terms, ϵ is most unlikely ever to be small enough to make σ/ϵ equal to a large number (cf. Table 1). It is therefore convenient to write

(6.6) $$\sigma = \epsilon E, \qquad E \leq \text{ord } 1,$$

with the order classes defined in the light of the limit $\sigma \to 0$.

It is helpful to disregard the diffusive or transport terms that contain Re, for the present, although Re *is* to be thought of as a large number throughout (Table 1).

It can now be seen that, although the *form* of the reaction terms will be different in both (5.10) and (5.11), depending on whether $E = \text{ord } 1$ or $E < \text{ord } 1$, they are always $O(\sigma_1 \sigma)$ in (5.10) and $O(\sigma_1 \sigma^2)$ in (5.11).

At this juncture one must make a very important decison about the physical character of the processes that are to be modelled, namely that *the gasdynamical and the chemical events* within the system *are to have equal status*. The left-hand side of (5.10) represents the gas dynamical changes that take place with time on wavelets of the constant-β family; in view of §6.1 this left-hand side must be $O(\sigma)$. In view of the fact, just described in the paragraph above, that the reaction terms in (5.10) are $O(\sigma_1 \sigma)$, it is clear that the chosen physical behaviour demands

(6.7) $$\sigma_1 = 1.$$

Observing the conditions in (5.12), and using the results in (6.1), (6.3),

(6.4) and (5.14) shows that the related value of ξ_β is

(6.8) $\quad \xi_\beta \sim \bar{a}_{f0}(\sigma\beta) + \frac{1}{2}(\gamma - 1)\sigma D'(\beta) - \frac{1}{2}(\gamma + 1) \int_{\sigma\beta}^{T} u_\beta^{(1)}(\beta, s) \, ds.$

This relation shows that ξ_β is $O(1)$, at least for $O(1)$ values of T and β. The fact that ξ_β may approach zero for some T in this range of values, if $u_\beta^{(1)}$ is positive, must be kept in mind. For the present, ξ_β equal to $O(1)$ is satisfactory and it is now possible to draw another important conclusion from the distinguished-limiting character of the present model.

From all of the information uncovered so far in this section it can now be seen that (5.11) demands that $\bar{q}$ shall behave like

(6.9) $\quad\quad\quad\quad\quad \bar{q} \sim \sigma^2 \, q^{(1)}(\beta, T).$

Only under these conditions can a balance exist between the (convective) terms on the left-hand side of (5.11) and the reaction terms on the right-hand side.

Thus, although the chemical reaction is to play a central role in the wave propagation process it does so with only small, second-order, changes in chemical composition *at this stage*.

6.3. Distinguished limits; Reynolds number. It is now necessary to consider the transport terms in the equations of §5.2.

With the small perturbation results established so far, it can be seen from (4.5) coupled with (6.1), (6.4) that

(6.10) $\quad\quad\quad\quad \bar{\theta} \sim \gamma \bar{p}_0 + \sigma \bar{a}_{f0}(\gamma - 1)u^{(1)}.$

Thus the transport terms in (5.10) become

(6.11) $\quad \dfrac{1}{\sigma \operatorname{Re} \xi_\beta} \left\{ \left[\dfrac{(\gamma - 1)\bar{\lambda}_0}{\operatorname{Pr}} + \bar{\eta}_0 \right] \bar{a}_{f0} \left(\dfrac{u_\beta^{(1)}}{\xi_\beta} \right)_\beta \right\} \equiv \dfrac{\bar{a}_{f0} \, \delta_0}{\sigma \operatorname{Re} \xi_\beta} \left(\dfrac{u_\beta^{(1)}}{\xi_\beta} \right)_\beta,$

to first order, where $\bar{\lambda}_0$ and $\bar{\eta}_0$ are the ambient atmosphere values of $\bar{\lambda}$ and $\bar{\eta}$ and hence dependent upon T and *not* upon β. In (5.11) the transport (diffusion) term is

(6.12) $\quad\quad\quad\quad\quad \dfrac{\sigma \overline{\mathcal{D}}_0}{\operatorname{Re} \operatorname{Sc} \xi_\beta} \left(\dfrac{q_\beta^{(1)}}{\xi_\beta} \right)_\beta,$

also to first order, and the viscous term in (5.9) is likewise

(6.13) $\quad\quad\quad\quad\quad \dfrac{\bar{\eta}_0}{\operatorname{Re}} \left(\dfrac{u_\beta^{(1)}}{\xi_\beta} \right)_\beta.$

The last term in the momentum equation has already been presumed to be negligible in the limit as $\sigma \to 0$, which clearly implies that $\operatorname{Re} \to \infty$ in some appropriate fashion.

In (5.10) the left-hand side is $O(\sigma)$, as is the reaction term, so that the transport terms (6.11) should only be retained if

(6.14) $$(\sigma^2 \text{ Re})^{-1} = O(1).$$

Since the left-hand side of (5.11) is $O(\sigma^2)$, as is the reaction term, it is clear that the diffusion term (6.12) should only be retained if

$$(\sigma \text{ Re})^{-1} = O(1).$$

Evidently (6.14) is the stronger condition on Re. When $\text{Re} = O(\sigma^{-2})$, the transport terms in (5.10) should therefore be retained, but they must necessarily be *excluded* from (5.9) and (5.11). A glance at Table 1 will indicate the order of size of σ implied by the requirement (6.14), since σ must be comparable with $\text{Re}^{-1/2}$. For additional comparison, σ lies roughly between 10^{-9} to 3×10^{-4} for acoustic signals audible to the human ear, so that signals for which transport effects are important are of audible levels for most of the activation energy range.

6.4. Small perturbation equations. As a result of the discussions of the earlier sections in this chapter one can now write down a consistent set of small perturbation equations for uni-directional wave propagation, with first order involvement of the combustion chemistry, as follows.

(6.15) $$\bar{a}_{f0}\rho^{(1)} = u^{(1)},$$

(6.16) $$\bar{a}_{f0}u^{(1)} = p^{(1)},$$

(6.17) $$p_T^{(1)} + \bar{a}_{f0}u_T^{(1)} = \frac{\epsilon}{\sigma} \exp\left\{\frac{1}{\epsilon}\left(1 - \frac{1}{\gamma \bar{p}_0}\right)\right\} \bar{q}_0^n$$
$$\times \left\{(1 + \sigma\rho^{(1)}) \exp\left(\frac{\sigma}{\epsilon} \frac{(\gamma - 1)}{\bar{a}_{f0}^3} u^{(1)}\right) - 1\right\}$$
$$+ \frac{\bar{a}_{f0}\delta_0}{\sigma^2 \text{ Re } \xi_\beta} \left(\frac{u_\beta^{(1)}}{\xi_\beta}\right)_\beta,$$

(6.18) $$\bar{a}_{f0}q_\beta^{(1)} \frac{1}{\xi_\beta} = -\frac{\epsilon}{\bar{Q}} \exp\left\{\frac{1}{\epsilon}\left(1 - \frac{1}{\gamma \bar{p}_0}\right)\right\} \bar{q}_0^n$$
$$\times \left\{\exp\left(\frac{\sigma}{\epsilon} \frac{(\gamma - 1)}{\bar{a}_{f0}^3} u^{(1)}\right) - 1\right\}.$$

Certain restrictions must be obeyed; they are

(6.19) $$\frac{\sigma}{\epsilon} = E \leq \text{ord } 1, \qquad \frac{1}{\sigma^2 \text{ Re}} \leq \text{ord } 1, \qquad \sigma \to 0.$$

These restrictions are a reiteration of (6.6) and a reinterpretation of the condition on Re derived in §6.3. Furthermore ξ_β must not be allowed to become too small without careful consideration of the consequences.

With these matters in mind, (6.15)–(6.18) provide a firm base for an attack on a whole spectrum of wave problems.

6.5. The ambient atmosphere: A restriction. There is a companion to the restrictions referred to in §6.4 that is sufficiently important to warrant a section to itself.

It must be observed here that the foregoing arguments are based on the tacit assumption that the coefficient function $\exp\{\epsilon^{-1}[1 - (\gamma\bar{p}_0)^{-1}]\}\bar{q}_0^n$ in (5.10) and (5.11) is $O(1)$ and, indeed, that $(1 + \bar{q}/\bar{q}_0)^n$ is likewise also $O(1)$.

If ϵ is small, it is clear that, as soon as the ambient atmosphere begins to change from the initial state $\gamma\bar{p}_0 = 1$ by an amount that exceeds $O(\epsilon)$, say to $\gamma\bar{p}_0$ equal to $(1 - \bar{B})^{-1}$, then the above coefficient function will become exponentially large like $\exp(\bar{B}/\epsilon)$, $\bar{B} > 0$. Consultation of (4.9) shows that this very swift rate of change in the ambient state will essentially occur before $\bar{q}_0$ diminishes significantly below the value unity. For the present, progress will be made on the continued understanding that the given coefficient function remains $O(1)$.

Since (6.18) shows that $q^{(1)}$ is bounded if ξ is bounded, which essentially means that the disturbance is of finite and $O(1)$ geometric extent, it can be seen that $(1 + \bar{q}/\bar{q}_0)^n = (1 + \sigma^2 q^{(1)}/\bar{q}_0)^n$ is properly behaved and the conditions described earlier for the validity of (6.15) to (6.18) are obeyed, at least in the early stages of the process. As will be described below it is necessary to exercise great care when the ambient reaction rate term $\exp\{\epsilon^{-1}[1 - (\gamma\bar{p}_0)^{-1}]\}\bar{q}_0^n$ begins to become large.

7. Linear chemistry. The equations in §6.4 are a suitably validated starting point for the examination of plane uni-directional waves, and explicit note can now be taken of the division between physical behaviours that is implied by (6.19). In particular, if

$$\sigma/\epsilon = E < \text{ord } 1,$$

the exponential terms in (6.17), (6.18) that involve this quotient must be rewritten so that the right-hand side of (6.17), for example, begins with the term

$$\exp\left[\frac{1}{\epsilon}\left(1 - \frac{1}{\gamma\bar{p}_0}\right)\right] \bar{q}_0^n \left\{\frac{(\gamma - 1)}{\bar{a}_{f0}^3} u^{(1)} + \epsilon\rho^{(1)}\right\}.$$

Thus this term, whose origin can be traced back from (6.17) through (5.10) and (4.4±) to (3.10±), and hence to the "reaction-rate difference" $\rho\mathcal{R} - \rho_i\mathcal{R}_0$ is linear in the perturbation quantities.

This case will be referred to from now on as the case of "linear chemistry".

Under this general heading there are a number of special circumstances, or distinguished limits, that deserve separate study.

7.1. General results for $E <$ ord 1. When E, defined in (6.19), approaches zero as $\sigma \to 0$, the equations in §6.4 combine and simplify to give

$$
\begin{aligned}
(7.1) \quad u_T^{(1)} = &\left\{ \tfrac{1}{2}[(\gamma - 1)\bar{a}_{f0}^4 + \epsilon \bar{a}_{f0}^2] \right. \\
&\left. \times \exp\left[\frac{1}{\epsilon}\left(1 - \frac{1}{\gamma \bar{p}_0}\right)\right] \bar{q}_0^n - \tfrac{1}{2}\bar{a}_{f0}^{-1}\bar{a}_{f0T} \right\} u^{(1)} \\
&+ (\delta_0/2\sigma^2 \, \mathrm{Re}\, \xi_\beta)(u_\beta^{(1)}/\xi_\beta)_\beta,
\end{aligned}
$$

$$
(7.2) \quad q_\beta^{(1)} = -[(\gamma - 1)\xi_\beta/\overline{Q}\bar{a}_{f0}^4]\exp\left[\frac{1}{\epsilon}\left(1 - \frac{1}{\gamma \bar{p}_0}\right)\right]\bar{q}_0^n u^{(1)}.
$$

It is necessary to retain the term $\epsilon \bar{a}_{f0}^2$ on the right-hand side of (7.1) since $\epsilon = O(1)$ is not excluded by the present form for E. Decisions about σ^2 Re are postponed for the moment.

Defining the coordinate Ξ via

$$
(7.3) \quad \sigma \Xi = \sigma \xi + \int_0^T \bar{a}_{f0}(s)\, ds - T
$$

allows (7.1) to be transformed into

$$
(7.4) \quad u_T^{(1)} - \tfrac{1}{2}(\gamma + 1)u^{(1)}u_\Xi^{(1)} = \{A(T) - \tfrac{1}{2}(\ln \bar{a}_{f0})_T\}u^{(1)} + \Delta(T)u_{\Xi\Xi}^{(1)},
$$

where $A(T)$ and $\Delta(T)$ are defined below:

$$
(7.5\mathrm{a}) \quad A(T) = \tfrac{1}{2}[(\gamma - 1)\bar{a}_{f0}^4 + \epsilon \bar{a}_{f0}^2]\exp\left[\frac{1}{\epsilon}\left(1 - \frac{1}{\gamma \bar{p}_0}\right)\right]\bar{q}_0^n
$$

$$
(7.5\mathrm{b}) \quad = \tfrac{1}{2}[(\gamma - 1)\bar{a}_{f0}^4 + \epsilon \bar{a}_{f0}^2]\frac{1}{\epsilon}\frac{d\bar{p}_0}{dT},
$$

$$
(7.6) \quad \Delta(T) = \tfrac{1}{2}\frac{\delta_0}{\sigma^2 \, \mathrm{Re}}.
$$

The second form of $A(T)$ in (7.5b) follows from (4.1d) and (4.10).

If (7.1), or (7.4), can be solved for $u^{(1)}$, (7.2) gives $q^{(1)}$ by simple quadrature. (7.4) is obviously very closely related to the Burgers equation and does, quite properly, reduce to it when the chemical term $A(T)$ is zero, and $\bar{a}_{f0}$ and $\Delta(T)$ are constants.

It is certainly correct here to think in terms of "short" time intervals $\gamma T \ll 1$, prior to the onset of the rapid explosive evolution of the ambient atmosphere, during which both $A(T)$ and $\Delta(T)$ are very nearly constant. Even under these simplified conditions there does not seem to be a linearising transformation of (7.4) equivalent to the Cole–Hopf reduction of the Burgers equation; indeed this particular reduction converts (7.4) into an equation of the form $\psi_T = \psi_{\Xi\Xi} + \psi \ln \psi$, which is certainly no simpler than the original equation, so that it is expedient to pursue the predictions of (7.1) or (7.4) through the medium of some additional ad hoc approximations.

7.2. Acoustical disturbances. It has already been remarked, at the end of §6.3, that audible acoustical disturbances have amplitudes of $\bar{u}$, for example, equal to 10^{-9} to 3×10^{-4}. If $\Delta(T)$ in (7.4) is to be $O(1)$ it is necessary to have σ^2 Re of this same order, and Table 1 shows that σ may therefore have to be too large for the disturbances to be acoustical. However, nothing that has been done so far is invalidated by the proposition that a function such as $u^{(1)}$ is itself small. For example one could make $u^{(1)}$ of the order of an amplitude number M, say, where M is very small, and in this way guarantee the acoustical character of the disturbances. The $O(1)$-character of σ^2 Re remains under these conditions, of course, and shows that acoustical disturbances can be affected by viscosity. In view of the role of $M\sigma$ as the indicator of amplitude and σ as the scale for variations with time (cf. (5.12) and note that we are now assuming that the function D' is of order M) it is clear that the viscous influences are strongly dependent on a typical frequency of the acoustic waves.

Assuming that the differentiation operations continue to be of $O(1)$ character, (7.4) can be linearised to give

$$u_T^{(1)} = A(T)u^{(1)} - \tfrac{1}{2}(\ln \bar{a}_{f0})_T u^{(1)} + \Delta(T)u_{\Xi\Xi}^{(1)}$$

or

(7.7) $\qquad \hat{u}_T^{(1)} = A(T)\hat{u}^{(1)} + \Delta(T)\hat{u}_{\Xi\Xi}^{(1)}, \qquad \hat{u}^{(1)} = \bar{a}_{f0}^{1/2} u^{(1)}.$

In the circumstances (7.7) describes the behaviour of acoustical disturbances that propagate along $\bar{x}$-positive under the combined influences of (linear) chemistry and diffusion.

If the induction time t_I is very long there will be a considerable interval of time during which both $A(T)$ and $\Delta(T)$ vary but little from their initial values. This is especially true if ϵ is very small. Noting that in such circumstances $\bar{a}_{f0}$ will also remain essentially equal to its initial value of unity, (7.3) shows that Ξ and ξ become synonymous with one another. Defining

(7.8) $\qquad\qquad A_i = A(0), \qquad \Delta_i = \Delta(0),$

it follows that (7.7) can then also be written in the form (note that $\hat{u}^{(1)} \simeq u^{(1)}$ since $\bar{a}_{f0} \simeq 1$ in the present approximation)

(7.9) $\qquad\qquad u_T^{(1)} + u_{\bar{x}}^{(1)} = A_i u^{(1)} + \sigma^2 \Delta_i u_{\bar{x}\bar{x}}^{(1)}.$

It must be noted that $\partial/\partial T$ and $\partial/\partial \bar{x}$ here are *not* $O(1)$ operations; they are in fact $O(1/\sigma)$ operations but (7.9) is convenient for the present. The boundary condition at the piston face, given in exact form by (5.2), can be simplified here to read

(7.10) $$u^{(1)}(0, T) = D'(T/\sigma),$$

where D' is a function to be chosen.

7.2(i). *A Laplace transform solution.* The solution of (7.9), (7.10) is most effectively approached through the use of Laplace transforms, since D' is always taken to be identically zero for all $T \leq 0$. It is useful to take a specific form for D' and the one chosen here is a sine function:

(7.11) $$u^{(1)}(0, T) = M \sin(\omega T/\sigma), \quad T > 0,$$

where ω is a constant that permits some variation in the frequency of the signal without tampering with the scaling parameter σ, and M is the small constant amplitude number that validates (7.7), or (7.9).

With the adoption of condition (7.11) Laplace transform theory shows that the solution for $u^{(1)}$ is given by

(7.12)
$$u^{(1)}(\bar{x}, T) = \frac{(\omega M/\sigma)}{2\pi i} \int_{Br} \exp\left\{sT + \frac{\bar{x}}{2\sigma^2 \Delta_i}[1 - (1 - 4A_i\Delta_i\sigma^2 + 4\Delta_i\sigma^2 s)^{1/2}]\right\} \times \frac{ds}{s^2 + (\omega/\sigma)^2}$$

where Br is the usual Bromwich contour. The elementary transformation

$$4\Delta_i\sigma^2 s + 1 - 4A_i\Delta_i\sigma^2 = z$$

allows one to rewrite (7.12) in the rather more convenient form

(7.13)
$$u^{(1)} = \exp\left\{A_i T - \frac{T}{4\Delta_i\sigma^2} + \frac{\bar{x}}{2\Delta_i\sigma^2}\right\}$$
$$\times \frac{4\Delta_i\sigma\omega}{2\pi i} \int_{Br} \frac{M \exp[(T/2\Delta_i\sigma^2)f(z, \eta)] \, dz}{(z - 1 + 4A_i\Delta_i\sigma^2)^2 + (4\Delta_i\omega\sigma)^2}$$

where

(7.14) $$f(z, \eta) = \tfrac{1}{2}z - \eta\sqrt{z}, \quad \eta = \bar{x}/T.$$

7.2(ii). *Evaluation by the method of steepest descents.* Since T and Δ_i are $O(1)$, while σ is small, the integral in (7.13) can be evaluated approximately by the method of steepest descents. The only saddle point for the function $f(z, \eta)$ for any given value of η is found at

(7.15) $$z = z_0 = \eta^2.$$

The steepest path through z_0 is found by equating the imaginary parts of $f(z, \eta)$ and $f(z_0, \eta)$ and so is given by

$$\text{Im}(\tfrac{1}{2}z - \eta\sqrt{z}) = 0.$$

With

$$z = r e^{i\vartheta}$$

this gives the polar form of the equation of the steepest path S, as follows

(7.16) $$\sqrt{r}\cos(\tfrac{1}{2}\vartheta) = \eta,$$

so that S is a large open loop contour that passes through z_0 and the two points $\pm i2\eta^2$ and ultimately, as $r \to \infty$, enfolds the whole negative axis. The only question that now arises in attempting to reconcile S with Br is whether or not the simple poles at

(7.17) $$z = z_{p\pm} = 1 - 4A_i\Delta_i\sigma^2 \pm i4\Delta_i\omega\sigma$$

lie within S or outside it. In the latter case $\int_{\text{Br}}$ is $\int_S$ plus the sum of the residues at $z_{p\pm}$; in the former case the two integrals are simply equal to one another. The condition for the poles to lie outside S is expressible as

(7.18) $$r_S < r_p$$

where r_S is r on S when ϑ has the value ϑ_p; from (7.17),

(7.19) $$\vartheta_p = \pm \tan^{-1}\{4\Delta_i\omega\sigma(1 - 4A_i\Delta_i\sigma^2)^{-1}\}.$$

Using (7.16) and (7.17), (7.18) can be translated into

(7.20) $$r_p + 1 - 4A_i\Delta_i\sigma^2 > 2\eta^2.$$

From the fact that σ is small [$A_i, \Delta_i = O(1)$], it is clear from (7.17) that $r_p \simeq 1$. Equation (7.20) therefore shows that great care must be exercised when η is in the neighbourhood of unity, since the poles will be very near to S and may lie inside it or outside it, depending on the detailed nature of all of A_i, Δ_i, σ and η.

It is clear that the combination of viscous-and-heat-conduction dissipation with the amplifying effects of the ambient chemistry leads to some complicated behaviour near the wave head, $\eta \simeq 1$, in present circumstances. It can be shown that there are some small departures of the signal head from its inert inviscid gas position at $\eta = 1$ as a result of diffusion Δ_i and chemical amplification A_i; there is therefore a modification of the ideal inviscid wavehead speed but since it differs from that particular value by an amount proportional to σ^2 it is not an effect of any detectable physical importance. Sufficiently far behind the $\eta = 1$ location $u^{(1)}$ is described to a high order of accuracy by the contributions from the simple poles in (7.13). After some simplification, based on $\sigma \to 0$ with A_i and Δ_i both $O(1)$, it transpires that

(7.21) $\qquad u^{(1)} \simeq \exp\{(A_i - \omega^2 \Delta_i)\bar{x}\} M \sin\left\{\dfrac{\omega}{\sigma}(T - \bar{x})\right\},$

provided that $\mathrm{ord}(T - \bar{x}) > \sigma$ (this last result comes from a closer analysis of the integral taken on the steepest path S).

7.2(iii). *Viscous damping and reaction-based amplification.* It is already well known that an exothermically disequilibrium atmosphere has an amplifying effect on any gas dynamic disturbances that propagate through it. Equation (7.21) confirms this result, but it also shows that the effect is linearly additive to the familiar frequency-dependent acoustic damping that is represented in (7.21) by the factor $\exp(-\omega^2 \Delta_i \bar{x})$. The form of the quantity can be traced back through the definitions in (7.6) and (6.11); some remarks about the size and physical character of the various terms will be made shortly, but meanwhile note condition (6.14), namely $\sigma \sim \mathrm{Re}^{-1/2}$. Remembering that M [see (7.11)] must be small, it follows that the present "acoustic" theory must be confined to amplitudes of disturbance very much less than $\mathrm{Re}^{-1/2}$ (cf. Table 1).

Equation (7.21) defines an initial zero growth frequency ω_{0i} as follows:

(7.22) $\qquad\qquad\qquad \omega_{0i} = (A_i/\Delta_i)^{1/2}.$

For all $\omega > \omega_{0i}$ harmonic disturbances start by being damped while for all $\omega < \omega_{0i}$ they begin by being amplified. It is interesting to note that this fact accounts for a good deal of the complicated behaviour near $\eta \simeq 1$ that has already been alluded to above. The value of ω_{0i} defined in (7.22) can be estimated once values of γ and Pr are chosen from which to calculate A_i and Δ_i. Equation (7.5b) shows that when $\epsilon \ll 1$

$$A_i \simeq \tfrac{1}{2}(\gamma - 1) = 0.2$$

when $\gamma = 1.4$; (7.6) with $\sigma^2 \mathrm{Re} = 1$ and (6.11) shows that

$$\Delta_i = \tfrac{1}{2}\{1 + (\gamma - 1)/\mathrm{Pr}\} = 0.78$$

when $\mathrm{Pr} = 0.72$. Thus

$$\omega_{0i} \simeq 0.51.$$

Of more practical interest is the actual frequency that this number represents; (7.21), for example, shows that in view of the nondimensionalizations in (4.1g) it must be given by

$$\omega_{0i}/\sigma \gamma t_l = \omega'_{0i} \approx \mathcal{P}\,\mathrm{Re}^{-1/2}/3.$$

The last result follows because σt_l is roughly equal to $\mathrm{Re}^{1/2}\mathcal{P}^{-1}$ when σ is $\mathrm{Re}^{-1/2}$ (see (4.14)). For the conditions appertaining to Table 1, ω'_{0i} is as shown in Table 2. For activation energies larger than about 30 the initial zero-growth

TABLE 2
Zero growth frequency ω'_{0i} (in Hertz) for plane acoustic waves for various values of activation energy number ϵ.

$1/\epsilon$	15	20	25	30	35	40	45
ω'_{0i}, Hz	3.7E + 7	1.1E + 6	1.0E + 5	9.0E + 3	8.0E + 2	7.0E + 1	6.0

frequency lies roughly in the audible range, while for smaller energies ω'_{0i} is in the ultrasonic range. What is perhaps of most interest is that ω'_{0i} appears to lie within the compass of practical acoustic frequencies. In other words, it is these frequencies that will be sustained in many explosive atmospheres. It is perhaps better to use sustained than amplified in this last sentence because the amount of possible amplification is necessarily limited by the requirement for the validity of (7.21). The dimensionless coordinate $\bar{x}$ is near to unity when an acoustic wave, starting at $\bar{x} = 0$, has travelled for a time of about one induction interval, t_I. Since (7.21) is limited to times well before t_I, it follows that $\bar{x}$ must be much less than unity for the present, and that reactive amplification of acoustical waves is quite small in these circumstances.

7.3. Nonlinear small disturbances. When σ is so large as to make σ^2 Re very large, the terms in (7.1) or (7.4) that are proportional to the diffusivity drop out and (7.1) simplifies to

(7.23) $$u_T^{(1)} = A(T)u^{(1)} - \tfrac{1}{2}(\ln a_{f0})_T u^{(1)},$$

where $\partial/\partial T$ is taken at fixed β and $A(T)$ is defined in (7.5a or b). Any σ values greater than or equal to (say) ten times $\text{Re}^{-1/2}$ (cf. Table 1) should suffice to make (7.23) a reasonable approximate version of (7.1), provided that ξ_β does not become too small. Alternatively one might decide that viscous effects will be postponed to the next order in the expansion of $\bar{u}$ as a series in powers of σ by making sure that $(\sigma^2 \text{Re})^{-1}$ is $O(\sigma)$; Table 1 shows that σ is roughly equal to 10^{-2} when $\epsilon = 1/20$, or 10^{-3} when $\epsilon = 1/30$. Under these conditions the "linear chemistry" model ($\sigma/\epsilon \ll 1$) is evidently satisfactory. What happens when ξ_β does approach zero will be discussed below.

The solution of (7.23), on the understanding that ξ_β does not misbehave, is

$$u^{(1)} = C(\beta)\bar{a}_{f0}^{1/2}(T) \exp\left\{\int_{\text{const.}}^T A(s)\, ds\right\},$$

where $C(\beta)$ is a function of integration that must be chosen to satisfy the boundary condition (5.12). Thus

(7.24) $$u^{(1)} = D'(\beta)\{f(T)/f(\sigma\beta)\}$$

where

(7.25) $$f(T) = \bar{a}_{f0}^{1/2}(T) \exp\left\{\int_0^T A(s)\, ds\right\}.$$

7.3(i). *The behaviour of* ξ_β. From (5.13), (6.3) and (6.4) it can be shown that

(7.26) $$\sigma[\xi + \sigma D(\beta)] = T - \int_{\sigma\beta}^T \bar{a}_{f0}(s)\, ds - \tfrac{1}{2}\sigma(\gamma + 1) \int_{\sigma\beta}^T u^{(1)}(\beta, s)\, ds,$$

or, in terms of Ξ defined in (7.3),

(7.27) $$\sigma[\Xi + \sigma D(\beta)] = \int_0^{\sigma\beta} \bar{a}_{f0}(s)\, ds - \tfrac{1}{2}\sigma(\gamma + 1) \int_{\sigma\beta}^T u^{(1)}(\beta, s)\, ds.$$

From (7.26) and (7.27), and using (5.12) to write $u^{(1)}(\beta, \sigma\beta)$ as $D'(\beta)$, it now follows that

$$\xi_\beta = \Xi_\beta = -\sigma D'(\beta) + \bar{a}_{f0}(\sigma\beta) + \tfrac{1}{2}(\gamma + 1)\sigma D'(\beta)$$

(7.28) $$- \tfrac{1}{2}(\gamma + 1) \int_{\sigma\beta}^T u_\beta^{(1)}(\beta, s)\, ds.$$

It is evidently only possible to find $\xi_\beta \to 0$ if $u_\beta^{(1)}$ is locally positive or, in other words, if the process is locally one of compression.

7.3(ii). *Compression waves and frozen chemistry.* Since

$$u_\Xi^{(1)} = u_\beta^{(1)}/\Xi_\beta$$

and $u_\beta^{(1)}$ is essentially $O(1)$, it follows that $u_\Xi^{(1)}$ increase in magnitude as Ξ_β diminishes. One can use (5.4) and (7.3) to show that

$$(u_T^{(1)})_\Xi = (u_T^{(1)})_\beta + (u_\beta^{(1)})_T (\beta_T)_\Xi$$

$$= (u_T^{(1)})_\beta + (u_\beta^{(1)})_T (\bar{a}_f - \bar{a}_{f0} + \bar{u})(\sigma\Xi_\beta)^{-1}.$$

It can now be seen that $(u_T^{(1)})_\Xi$ must also grow as Ξ_β diminishes because $(u_\beta^{(1)})_T$ is at least $O(1)$, as can be seen from (7.24).

Thus in compressive parts of the disturbance each of the first, second and fourth terms in (7.4) begins to grow in magnitude in a way that the third (ambient reaction-induced) term does not. To begin with, the fact that Δ is $o(1)$ in the present case means that *both* right-hand side terms in (7.4) play a minor role, and local $u^{(1)}$ changes approximately follow the simple equation

$$u_T^{(1)} - \tfrac{1}{2}(\gamma + 1)u^{(1)}u_\Xi^{(1)} \approx 0$$

that describes the progress of undiffused weak nonlinear waves in a chemically inert atmosphere. As the value of both $u_T^{(1)}$ and $u_\Xi^{(1)}$ continues to increase, there will come a time when $\Delta u_{\Xi\Xi}^{(1)}$ is of the same order as these changes. Then (7.4) goes over to the familiar Burgers equation and subsequent local motions must follow its behaviour patterns, which are essentially those of a locally chemically inert but now diffused flow. Thus, not unexpectedly, the compressions (or shocks) that appear in the present situation are of the chemically frozen kind. Of course the details of the transition from an initially smooth compression, appropriately spread out in space, to the local shock wave configuration is quite complicated and really demands the solution of (7.4) complete. When the thicknesses of local regions of compression, over which viscous effects (last term in (7.4)) are no longer negligible, are themselves small compared with the spatial extent of the whole disturbance field the shocks can be treated as discontinuities.

7.3(iii). *Weak shocks as discontinuities.* A relation that enables one to fit shocks treated as discontinuities into otherwise continuous flow fields can be derived from (7.4) as follows. First integrate (7.4) with respect to Ξ from Ξ_{s-} to Ξ_{s+}, where the latter are values of Ξ just behind and just ahead of an assumed shock discontinuity (they are therefore functions of time T). The result is

$$(7.29) \quad \int_{\Xi_{s-}}^{\Xi_{s+}} u_T^{(1)}(T, \hat{\Xi}) \, d\hat{\Xi} - \tfrac{1}{4}(\gamma + 1)[u^{(1)2}]_-^+$$
$$= A(T) \int_{\Xi_{s-}}^{\Xi_{s+}} u^{(1)}(T, \hat{\Xi}) \, d\hat{\Xi} + \Delta(T)[u_\Xi^{(1)}]_-^+,$$

where $[y]_-^+$ represents the difference between values of y at Ξ_{s+} and Ξ_{s-}.

The discontinuity is modelled by letting $\Xi_{s-} \to \Xi_{s+} \to \Xi_s$. Since $u^{(1)}$ is continuous, the integral on the right-hand side of (7.29) vanishes in the limit. Ahead of and behind the compression $u_\Xi^{(1)}$ is $O(1)$, so that the last term on the right-hand side of (7.29) vanishes because $\Delta(T)$ is $o(1)$[1]. The general and exact result,

$$(7.30) \quad \frac{d}{dT}\left\{\int_{\Xi_{s-}}^{\Xi_{s+}} u^{(1)}(T, \hat{\Xi}) \, d\hat{\Xi}\right\}$$
$$= u_+^{(1)} \frac{d\Xi_{s+}}{dT} - u_-^{(1)} \frac{d\Xi_{s-}}{dT} + \int_{\Xi_{s-}}^{\Xi_{s+}} u_T^{(1)}(T, \hat{\Xi}) \, d\hat{\Xi},$$

in an obvious notation, can be used to eliminate the first integral from (7.29) since the first integral in (7.30) must vanish as $\Xi_{s-} \to \Xi_{s+} \to \Xi_s$ from the continuity of $u^{(1)}$. Thus (7.29) becomes

$$-\frac{d\Xi_s}{dT}[u^{(1)}]_-^+ - \tfrac{1}{4}(\gamma + 1)[u^{(1)2}]_-^+ = 0$$

---

[1] In this section.

or, using (7.3) and (5.3),

$$\frac{d\bar{x}_s}{dT} = \bar{a}_{f0}(T) + \tfrac{1}{4}(\gamma + 1)\sigma(u_+^{(1)} + u_-^{(1)}) \tag{7.31}$$

where $u_+^{(1)}$, $u_-^{(1)}$ are the values of $u^{(1)}$ ahead of and behind the wave. This relation shows that the discontinuity is a (weak) frozen Rankine–Hugoniot shock. Further examination of nonlinear small disturbance behaviour will be on the assumption that the discontinuous shock description is of adequate accuracy.

7.3(iv). *A weak shock fitting formula.* It is necessary to go further with the question of how to fit a shock or shocks into the system so as to satisfy the basic weak shock formula given in (7.31). First note from (7.4) and the relation

$$\sigma\xi = T - \bar{x}$$

between ξ, T and $\bar{x}$ that

$$\bar{x} = \sigma^2 D(\beta) + \int_{\sigma\beta}^{T} \{\bar{a}_{f0}(s) + \tfrac{1}{2}\sigma(\gamma + 1)u^{(1)}(\beta, s)\}\, ds. \tag{7.32}$$

Therefore

$$(\bar{x}_T)_\beta = \bar{a}_{f0}(T) + \tfrac{1}{2}\sigma(\gamma + 1)u^{(1)}(\beta, T). \tag{7.33}$$

It is convenient to rewrite (7.32) in a form which takes due note of (7.24) and (7.25) and makes use of the function

$$g(\beta) = D'(\beta)/f(\sigma\beta). \tag{7.34}$$

Thus (7.32) becomes

$$\bar{x} = -B(\beta) + \int_0^T \bar{a}_{f0}(s)\, ds + \tfrac{1}{2}\sigma(\gamma + 1)g(\beta)\int_0^T f(s)\, ds, \tag{7.35}$$

where the function $B(\beta)$ is defined as follows,

$$B(\beta) = \int_0^{\sigma\beta} \bar{a}_{f0}(s)\, ds + \tfrac{1}{2}\sigma(\gamma + 1)g(\beta)\int_0^{\sigma\beta} f(s)\, ds - \sigma^2 D(\beta). \tag{7.36}$$

It can be seen from (7.31) and (7.33) that

$$2\frac{d\bar{x}_s}{dT} \simeq (\bar{x}_T)_{\beta+} + (\bar{x}_T)_{\beta-}. \tag{7.37a}$$

Now $d\bar{x}_s/dT$ is the rate of change of $\bar{x}$ on the shock and can therefore be evaluated at *either* β_+, or β_-. Thus, more symmetrically, (7.37a) requires

$$\left(\frac{d\bar{x}_s}{dT} - \bar{x}_T\right)_{\beta+} + \left(\frac{d\bar{x}_s}{dT} - \bar{x}_T\right)_{\beta-} = 0. \tag{7.37b}$$

But (7.35) shows that when β is a function of time T, it is always true that

$$\frac{d\bar{x}}{dT} - \bar{x}_T = -B'(\beta)\dot{\beta} + \tfrac{1}{2}(\gamma + 1)g'(\beta)\sigma\dot{\beta}\int_0^T f(s)\,ds,$$

where B' and g' are the β-derivatives of B and g, and $\dot{\beta}$ is the T-derivative of β. Eliminating $\int_0^T f(s)\,ds$ by using (7.35) to equate $\bar{x}_s(\beta = \beta_+)$ and $\bar{x}_s(\beta = \beta_-)$ leads to the following relation between β_+ and β_- on the shock.

$$\dot{\beta}_+\{[B(\beta_-) - B(\beta_+)]g'(\beta_+) - B'(\beta_+)[g(\beta_-) - g(\beta_+)]\}$$
$$+\, \dot{\beta}_-\{[B(\beta_-) - B(\beta_+)]g'(\beta_-) - B'(\beta_-)[g(\beta_-) - g(\beta_+)]\} = 0,$$

which can be integrated to give

(7.38) $\qquad [B(\beta_-) - B(\beta_+)][g(\beta_-) + g(\beta_+)] = 2\int_{\beta_+}^{\beta_-} g(z)B'(z)\,dz.$

This shock fitting rule has a formal similarity with the "equal-areas" formula, that arises in the dynamics of inert gases, but the variable property background and the general amplification of disturbances due to their interaction with the ambient chemistry gives rise to some interesting modulations in the present case. Some simplification of (7.38) can be achieved by dropping the terms from (7.36) that are $O(\sigma^2)$, so that

(7.39) $\qquad\qquad B(\beta) \simeq \int_0^{\sigma\beta} \bar{a}_{f0}(s)\,ds.$

The $O(\sigma^2)$ terms arise from the fact that the boundary condition has actually been satisfied exactly in the solutions so far.

7.3(v). *Shock formation in a short compression pulse.* It is now possible to illustrate the behaviour of weak nonlinear waves in the presence of linear chemistry by working briefly through some examples. Consider first the particular case of a piston input for which

(7.40) $\qquad\qquad D'(\beta) = \begin{cases} M\sin\omega\beta, & 0 \leq \beta \leq \pi/\omega, \\ 0, & \pi/\omega < \beta, \end{cases}$

where M, ω are positive order one constants. This contrasts with (7.11) where M is small; also the present input is for a limited duration only. In order to use (7.38) to fit a shock wave into this single sine-hump disturbance, note that the function $f(\sigma\beta)$, that appears in (7.34) defining $g(\beta)$, and which is itself defined in (7.25), is equal to unity for all practical purposes since $D'(\beta)$ vanishes when $\beta > \pi/\omega$ and the value of $f(\sigma\beta)$ is thus only required in the interval of time from O to $\sigma\pi/\omega$; as $\sigma \to 0$, $f(\sigma\beta) \to 1$ for order one values of β. Since $\bar{a}_{f0}(s)$ is approximately unity for small values of s, it follows from (7.39) that

(7.41) $$B(\beta) \simeq \sigma\beta$$

is adequate in present circumstances.

A shock will form where ξ_β first vanishes or, equivalently, where T has the least value compatible with the condition

$$\bar{x}_\beta \simeq -\sigma + \tfrac{1}{2}\sigma(\gamma + 1)D''(\beta)\int_0^{T_{\min}} f(s)\,ds = 0,$$

as can be seen by noting that $\sigma\xi = T - \bar{x}$ and using the above approximations in (7.35). Thus $T_{\min}$ is found on the leading wavelet $\beta = 0$ and has the value

(7.42) $$T_{\min} \approx 2/(\gamma + 1)M\omega.$$

The requirement for $T_{\min}$ to lie well within the induction interval can be written as follows

(7.43) $$\frac{2}{(\gamma + 1)M\omega} \ll \frac{1}{\gamma} \quad \text{or} \quad \frac{1}{\omega} \ll \frac{1}{2}\frac{(\gamma + 1)}{\gamma}M,$$

since $1/\gamma$ is the value of T at the end of this interval. On the piston $T = \sigma\beta$ so that (7.40) requires $0 \le T \le \pi\sigma/\omega$ or, in dimensional terms (see (4.1)), $0 \le t \le \pi\sigma\gamma t_I/\omega$. Hence the pulse duration is $\pi\sigma\gamma t_I/\omega$ and if (7.43) is to be satisfied it is necessary to have

(7.44) $$\frac{\pi\sigma\gamma t_I}{\omega} \ll \tfrac{1}{2}(\gamma + 1)\pi\sigma t_I M.$$

Assuming that σ is indeed equal to $10\,\text{Re}^{-1/2}$, as proposed at the start of §7.3, it follows from Table 1 that σt_I has the values exhibited in Table 3, which also lists the relevant σ values as well.

TABLE 3
Order of magnitude of pulse duration σt_I (in seconds) for a shock to form within the induction interval, and asociated pulse amplitude, for various values of activation energy number ϵ.

$1/\epsilon$	15	20	25	30	35	40	45
σt_I, sec	1.3E − 6	4.4E − 6	4.8E − 5	5.4E − 4	6.1E − 3	6.9E − 2	7.8E − 1
σ	7.4E − 1	2.2E − 3	2.0E − 3	1.8E − 4	1.6E − 6	1.4E − 7	1.2E − 8

Amplitudes as high as 0.74 cannot be considered small so that there is a lower limit on $1/\epsilon$ below which the present theory cannot be valid. If $1/\epsilon$ exceeds 20, the basic theory can be expected to behave itself but it is clear from (22) and Table 3 that very short, microsecond, pulse durations will be required before shock waves are formed at the pulse head of compressions like the one exemplified in (7.40).

7.3(vi). *Evolution of the weak shock wave.* Assuming that the conditions for $M\omega$ described in (7.43) are met, the shock fitting condition (7.38) gives

$$(7.45) \qquad (\beta_- - \beta_+) \sin \omega\beta_- = \frac{2}{\omega}(1 - \cos \omega\beta_-),$$

for the short pulse defined in (7.40). This relatively simple result is derived because $g(\beta) \simeq D'(\beta)$ is identically zero for all $\beta < 0$, and hence for all β_+ values. Using (7.35) to equate $\bar{x} - \int_0^T \bar{a}_{f0}(s)\,ds$ on the shock wave when β is equal either to β_+ or to β_- gives

$$(7.46) \qquad \beta_- - \beta_+ \simeq \tfrac{1}{2}(\gamma + 1)M \sin \omega\beta_- \int_0^T f(s)\,ds,$$

where β_- lies in the range $0 \leq \beta_- \leq \pi/\omega$ and (7.41) has been used for $B(\beta_-)$. Combining (7.45) and (7.46) shows that

$$\tfrac{1}{2}(\gamma + 1)M \sin^2 \omega\beta_- \int_0^T f(s)\,ds = \frac{2}{\omega}(1 - \cos \omega\beta_-)$$

from which it follows that

$$(7.47) \qquad \cos \omega\beta_- = \left\{\tfrac{1}{4}(\gamma + 1)M\omega \int_0^T f(s)\,ds\right\}^{-1} - 1.$$

Equation (7.47) shows that $0 \leq \beta_- < \pi/\omega$ for $T_{\min} \leq T < \infty$. In practice the maximum value for T allowed by the present theory is $1/\gamma$ to a first order, since we are only modelling events *before* the onset of homogeneous explosion.

The value of the velocity $u_-^{(1)}$ just behind the shock is given by $u^{(1)}[\beta_-(T)]$; since (7.24) gives

$$u_-^{(1)} = u^{(1)}[\beta_-(T)] \simeq f(T)M \sin \omega\beta_-$$

with sufficient accuracy, it follows from (7.47) that

$$(7.48) \qquad u_-^{(1)} \simeq f(T)M \left\{1 - \left[1 - \left(\tfrac{1}{4}(\gamma + 1)M\omega \int_0^T f(s)\,ds\right)^{-1}\right]^2\right\}^{1/2},$$

where $f(T)$ is defined in (7.25). Since $u_+^{(1)} \equiv 0$ (7.48) gives the shock strength, the shock path $\bar{x}_s(T)$ follows from (7.31) with these values of $u_+^{(1)}$ and $u_-^{(1)}$. It must be remembered that (7.48) is only valid for T in the interval $T_{\min} \leq T < 1/\gamma$ (see (7.42)).

7.3(vii). *Changes in chemical composition.* Equations (7.2) and (7.5) can be combined together with the fact that $\bar{a}_{f0}^2$ is equal to $\gamma\bar{p}_0$ to show that

$$(7.49) \qquad (\overline{Q}/\xi_\text{B})q_\text{B}^{(1)} = -2\left[1 + \frac{\epsilon}{(\gamma - 1)}\bar{p}_0\right]^{-1} A(T)u^{(1)} \simeq -2A(T)u^{(1)}.$$

The approximate final result here is tolerable when ϵ is small, which on the whole it is, and certainly makes the following analysis more concise. Since

$$\frac{1}{\xi_\beta} q_\beta^{(1)} = q_\xi^{(1)} = -\sigma q_{\hat{x}}^{(1)}$$

(7.49) shows that

(7.50) $$\overline{Q}\sigma q^{(1)} \simeq 2A(T) \int_{L(T)}^{\bar{x}} u^{(1)} \, d\bar{x} + H(T),$$

where $H(T)$ is an arbitrary function of T at this stage, as is $L(T)$.

The disturbances are not created until $T = 0$, which implies that the input-function $D'(\beta) \equiv 0$ for all $\beta \leq 0$. Recalling from (5.3) that $\sigma\xi$ is equal to $T - \bar{x}$ (since σ_1 in (5.1) is necessarily unity; see §6.2, especially (6.7)), (7.26) confirms the intuitive result that the wavelet $\beta = 0$ follows a path

(7.51) $$\bar{x} = \bar{x}_h(T) = \int_0^T \bar{a}_{f0}(s) \, ds,$$

since $u^{(1)}(0, T)$ vanishes by hypothesis. For any $\bar{x} > \bar{x}_h(T)$, $q^{(1)}$ must be zero and (7.50) therefore gives

(7.52) $$\overline{Q}\sigma q^{(1)} \simeq -2A(T) \int_{\bar{x}}^{\bar{x}_h(T)} u^{(1)}(T, \hat{x}) \, d\hat{x}, \quad \bar{x} < \bar{x}_h(T).$$

The result in (7.52) is only valid until shocks form in the disturbance. If, as in the example discussed in §§7.3(v) and (vi), the shock is at the head of the disturbance, it is only necessary to replace $\bar{x}_h(T)$ in (7.52) by $\bar{x}_s(T)$, calculated from (7.31). The formal change that this makes in (7.52) is not significant; more complication may ensue if there are several shock waves in the disturbance but it is not difficult to write down a general result in light of the fact that $q^{(1)}$ is continuous across any shock wave (§§7.3(ii) and (iii)) while $u^{(1)}$ jumps in value.

Although (7.52), or any of its modified versions to account for the presence of shocks, is an innocuous looking result it nevertheless has some profound implications. If the right-hand side of (7.52) is not limited to $O(\sigma)$ in magnitude, $q^{(1)}$ is not $O(1)$, and the whole of the present theory must fail. Such misbehaviour of (7.52) comes, in essence, from two sources.

a) The integral of $u^{(1)}$ between $\bar{x}$ and $\bar{x}_h(T)$ or $\bar{x}_s(T)$ may exceed $O(\sigma)$ (assume that $A(T)$ is $O(1)$, for the present). For example $u^{(1)}$ may be positive everywhere (it is essentially $O(1)$) between $\bar{x} = 0$ and $\bar{x}_h$; then the present theory can only be valid for times T that are themselves $O(\sigma)$. This situation occurs if the input disturbance is given by

$$D'(\beta) = \begin{cases} 0, & \beta < 0, \\ 1, & \beta > 0, \end{cases}$$

which represents the instantaneous establishment of a (weak) shock wave, followed by maintenance of the compression by continued movement of the piston. The sustained higher rate of chemical energy release at the piston-face

$$\bar{x} = \bar{x}_p(T) = \sigma T$$

under these conditions leads to values of $q^{(1)}$ that exceed $O(1)$ for any times T that exceed $O(\sigma)$ (remember that $\bar{x}_h(T)$ is roughly equal to T, since $\bar{a}_{f0}$ is unity for all practical purposes for small T values). It will be necessary to keep a careful watch on the order of magnitude of the operator $\partial/\partial\beta$ at fixed T under conditions that may lead to the present form of behaviour.

b) Even though the integral in (7.52) may never exceed $O(\sigma)$, as in the case of the short compression pulse described in §7.3(vi) (cf. (7.26) and (7.40)), conditions in the ambient atmosphere may advance to the point that $A(T)$ will become very large.

In either case, a) or b), revision of the present theory is necessary. The matter is of sufficient interest to warrant examination in a separate section.

To close this present section note that (7.49) integrates directly to give

$$\bar{Q}q^{(1)} \simeq -2A(T) \int_0^\beta \xi_\beta u^{(1)} \, \partial\beta \qquad (7.53)$$

(the lower limit may be replaced by $\beta_-(T)$ if there is a shock at the head of the disturbance). (5.12) reminds us that β may exceed $O(1)$ if T is greater than σ in order of magnitude, whence the possible failure of $q^{(1)} = O(1)$ described in a) is reiterated. More importantly at this juncture it is observed that, while $(q_\beta^{(1)})_T$ from (7.53) remains $O(1)$, $(q_T^{(1)})_\beta$ *does not* do so in case a).

Thus not only does $\bar{q}$ grow to magnitudes greater than $O(\sigma^2)$ but $(\bar{q}_T)_\beta$ begins to exceed $(\bar{q}_\beta)_T$. The next phase of wave propagation evidently requires, not only a revision of the amplitudes of chemical composition variation, but a reconsideration of rates of change in general (i.e. with respect to time and space).

8. Breakdown of uni-directional propagation. Attention will be confined entirely to the case for which σ is large enough to make viscous effects insignificant away from the interior of shock waves. The condition $E <$ ord 1 (see §6.2) will also be adhered to, so that it is essentially the "linear chemistry" situation described in §7.3 that will form the basis of the present discussion.

8.1. Behaviour of functions $A(T)$ and $f(T)$. The amplitude factor $A(T)$ is defined in (7.5), and is repeated here for convenience (note that $\bar{a}_{f0}^2 = \gamma\bar{p}_0$), with the additional assumption that the terms proportional to ϵ can be neglected; then

(8.1a) $\quad A(T) = [(\gamma - 1)/2\gamma^2 \bar{p}_0^2] \exp \left\{ \dfrac{1}{\epsilon} \left(1 - \dfrac{1}{\gamma \bar{p}_0} \right) \right\} \bar{q}_0^n,$

(8.1b) $\quad\quad\quad = -[(\gamma - 1)/2\gamma\epsilon] \dfrac{d}{dT} (1/\gamma \bar{p}_0).$

The second form (8.1b) follows from (4.10) and shows at once that

$$\int_0^T A(s)\, ds = [(\gamma - 1)/2\gamma\epsilon] \left(1 - \dfrac{1}{\gamma \bar{p}_0} \right).$$

The amplification factor defined in (7.25) is therefore given by

(8.2) $\quad f(T) = (\gamma \bar{p}_0)^{-1/2} \exp \left\{ [(\gamma - 1)/2\gamma\epsilon] \left(1 - \dfrac{1}{\gamma \bar{p}_0} \right) \right\}.$

It is helpful to define

(8.3) $$1 - \dfrac{1}{\gamma \bar{p}_0} = \bar{B}$$

so that (8.1a) and (8.2) can be rewritten in the forms

(8.4) $\quad A(T) = \tfrac{1}{2}(\gamma - 1)(1 - \bar{B})^2 \bar{q}_0^n e^{\bar{B}/\epsilon},$

(8.5) $\quad f(T) = (1 - \bar{B})^{1/2} \exp\{(\gamma - 1)\bar{B}/2\gamma\epsilon\}.$

A useful expression for $\bar{q}_0$ derives directly from (4.9) and (8.3), namely

(8.6) $\quad \bar{q}_0 = (\gamma \bar{p}_{0m} - \gamma \bar{p}_0)/\gamma \bar{Q} = (\bar{B}_m - \bar{B})/(1 - \bar{B}_m)(1 - \bar{B})\gamma \bar{Q},$

where

(8.7) $$\bar{B}_m = 1 - \dfrac{1}{\gamma \bar{p}_{0m}}.$$

Using the conditions appropriate to Table 1 for purposes of illustration, i.e. $\bar{Q} = \bar{p}_{0m} - 1 = 4$, $n = 1$, $\gamma = 1.4$, so that $\bar{B}_m = \tfrac{6}{7}$, and choosing the case $\epsilon = \tfrac{1}{30}$ the values of $\bar{B}(T)$, $A(T)$ and $f(T)$ are related as shown in Table 4. Clearly $f(T)$ will lead to significant growth of $u^{(1)}$ (cf. (7.24)) but the ultra-rapid growth of $A(T)$, that occurs after $\bar{B}(T)$ exceeds about 0.1 in the illustrative situation in Table 4, will have a dominant part to play in time intervals near to t_I. This is the situation described in §7.3(vii b).

8.2. Behaviour as $A(T)$ becomes large. When $A(T)$ is large, (7.23) makes it clear that

$$(u_T^{(1)})_B \sim A(T) u^{(1)}.$$

TABLE 4
Values of the functions $A(T)$ and $f(T)$ for various ambient pressure values given by $\bar{B}$ (see 8.3)).

$\bar{B}(T)$	$A(T)$	$f(T)$
0	0.2	1
0.1	3.4	1.46
0.2	5.30E + 1	2.11
0.3	7.90E + 2	3.03
0.4	1.1 E + 4	4.30
0.5	1.46E + 5	6.03
0.6	1.69E + 6	8.28
0.7	1.55E + 7	11.0
0.8	7.67E + 7	13.79
0.85	3.18E + 7	14.8
0.855	1.07E + 7	14.86
0.856	5.85E + 6	14.87
0.857	7.48E + 5	14.89
6/7	0	14.89

The solution (7.24) can be used to show that in similar circumstances

$$(\bar{u}_\beta)_T \sim \{D''(\beta)/D'(\beta)\}\bar{u} - \sigma A(\sigma\beta)\bar{u}.$$

Since the maximum value of $\sigma\beta$ is T, achieved on the piston face (see (5.12)), $\bar{u}_\beta$ is always less than $\bar{u}_T$ in the present situation when $\bar{B}$ is less than about 0.8.

Note that the supposition that $A(T)$ is large means that the disturbance has been travelling for a time approaching the induction time t_I, or $T \rightarrow 1/\gamma$ in dimensionless coordinates. From the work of §7.3(vii) it is clear that the disturbance must either be of the "short pulse" variety, as in §7.3(v), or the integrals (7.52) or (7.53) are bounded by σ or unity, respectively, as a result of alternations in the sign of $u^{(1)}$; if this is not so the type of breakdown described in §7.3(vii a) will have occurred before the present type b) failure.

The equations derived in §6 on the strict proposition that $(\partial/\partial T)_\beta$ is a $O(1)$ operation must begin to fail when $A(T)$ becomes of order $1/\sigma$. If, as suggested in §7.3, σ is roughly ten or so times the $Re^{-1/2}$ value listed in Table 1, that is to say about 1.8×10^{-4} in the present case, for which $1/\epsilon = 30$, this value of $A(T)$ is achieved when the ambient atmosphere has advanced to a state for which the $\bar{B}$ value lies between 0.3 and 0.4 (see Table 4).

It can be seen from (8.4), (8.6) that $A(T)$ is $O(1/\sigma)$ when

$$\bar{B} = O(\epsilon \ln \sigma).$$

It is therefore convenient to write

(8.8) $$\bar{B} = -\epsilon \ln \sigma + \epsilon B^{(1)},$$

where $B^{(1)}$ is $O(1)$ by hypothesis. Note that if $\overline{B}$ is to have a value, b say, that is essentially of order 0.3 to 0.4 in magnitude then ϵ must be given by

(8.9) $$\epsilon = -b/\ln \sigma = o(1).$$

This newly-acquired distinguished limit (cf. §§6.2 and 6.3) is consistent with the basic requirement that

$$E = \sigma/\epsilon = (-\sigma \ln \sigma)/b = o(1).$$

It now follows from (4.10) and the fact that $\epsilon \to 0$ as $\sigma \to 0$ (see (8.9)) that $B^{(1)}$ satisfies the equation

(8.10) $$-\frac{dB^{(1)}}{d\tau} = \hat{b} e^{B^{(1)}},$$

where τ is defined so that

(8.11) $$\sigma \frac{d\tau}{dT} = 1,$$

and the $O(1)$ constant quantity $\hat{b}$ is defined to be

(8.12) $$\hat{b} = [\gamma/(\gamma \overline{Q})^n](1-b)^2\{(1-\overline{B}_m)^{-1} - (1-b)^{-1}\}^n.$$

Equation (8.11) introduces a new time variable τ such that

(8.13) $$\sigma\tau = T + \text{constant} = T - T_b,$$

and it is sensible to choose T_b as the time at which $\overline{B}$ is equal to b. In view of the definition of T_b in (8.13) it follows that $B^{(1)}(0)$ is zero; this provides an initial condition for (8.10), which therefore makes

(8.14) $$B^{(1)}(\tau) = -\ln(1 - \hat{b}\tau).$$

8.3. Derivation of new variables. Although it is introduced in §8.2 as a time variable suitable for a description of ambient atmosphere behaviour in the neighbourhood of the time at which $\overline{B}$ has the order unity value b, it is evidently important to use τ to describe *disturbance* behaviour in this same neighbourhood. There is no need at this juncture to make any adjustments in the wavelet (roughly, the spatial) coordinate, since $(\overline{u}_\beta)_T$ is still well behaved (see §8.2).

So far the gas velocity has been estimated from the asymptotic representation

$$\overline{u} \sim \sigma u^{(1)}(T, \beta)$$

under the limiting process $\sigma \to 0$; β, T fixed. With the breakdown of this representation it is now proposed to try

FINITE AMPLITUDE WAVES IN COMBUSTIBLE GASES 221

(8.15) $$\bar{u} \sim G(\sigma)u_e^{(1)}(\tau, \beta)$$

with the limiting process $\sigma \to 0$; τ, β fixed. In terms of the intermediate time variable τ_i, defined by

(8.16) $\quad \sigma\tau = \delta(\sigma)\tau_i = T - T_b, \quad \delta(\sigma) = o(1), \quad \sigma/\delta(\sigma) = o(1),$

the new representation can be matched to the old one by requiring that the difference

(8.17) $\quad \sigma u^{(1)}[T_b + \delta(\sigma)\tau_i, \beta] - G(\sigma)u_e^{(1)}[\tau_i\delta(\sigma)/\sigma, \beta]$

will approach zero, at least to leading order, as $\sigma \to 0$; τ_i, β fixed. It is therefore necessary to make

(8.18) $\quad G(\sigma) = \sigma, \quad u_e^{(1)}(\tau \to -\infty, \beta) \to u^{(1)}(T_b, \beta) = D'(\beta)f(T_b)$

since $T < T_b$ in the overlap domain between the new and the old time regions. The "short pulse" approximation has been used to write in this last result (cf. (7.24)).

The continuity and momentum equations (5.8), (5.9) require that pressure and density variables shall be of the same order as variations in velocity, so that new representations for $\bar{p}$ and $\bar{\rho}$ are taken in the forms

(8.19) $\quad \bar{p} \sim \sigma p_e^{(1)}(\tau, \beta), \quad \bar{\rho} \sim \sigma \rho_e^{(1)}(\tau, \beta).$

Together with (8.15), (8.18) for $\bar{u}$, these representations need only be supplemented by the estimate

(8.20) $$\bar{q} \sim \sigma q_e^{(1)}(\tau, \beta)$$

whose form is made evident by the foregoing discussion and the solutions in §7.3(vii), in order to begin the analysis of the next phase of development of the disturbance.

From the fact that

(8.21) $$\begin{aligned} -\xi_\beta &= \frac{1}{\sigma}\bar{x}_\beta = \sigma D'(\beta) - \bar{a}_{f0}(\sigma\beta) - \tfrac{1}{2}\sigma(\gamma + 1)u^{(1)}(\beta, \sigma\beta) \\ &\quad + \int_{\sigma\beta}^{T} \tfrac{1}{2}(\gamma + 1)u_\beta^{(1)}(\beta, s)\, ds \\ &= -1 + \tfrac{1}{2}(\gamma + 1)\int_0^T u_\beta^{(1)}(\beta, s)\, ds + O(\sigma), \end{aligned}$$

see (7.28), it can be seen that ξ_β is $O(1)$.

Using the new variables and representations in the exact equations of §5 leads to the following equations for $u_e^{(1)}$, $p_e^{(1)}$, $\rho_e^{(1)}$ and $q_e^{(1)}$. (5.8) gives (NB $\sigma_1 = 1$ according to (6.7))

(8.22) $\quad \xi_\beta \rho_{e\tau}^{(1)} + \bar{a}_f(T_b + \sigma\tau)\rho_{e\beta}^{(1)} - u_{e\beta}^{(1)} = 0,$

while (5.9) gives

(8.23) $$\xi_\beta u^{(1)}_{e\tau} + \bar{a}_f(T_b + \sigma\tau)u^{(1)}_{e\beta} - p^{(1)}_{e\beta} = 0,$$

and one must remember that it has been presumed that $(1/\sigma^2 \text{ Re}) \to 0$ as $\sigma \to 0$.

Then (5.10), (5.11) show that

(8.24) $$p^{(1)}_{e\tau} + \bar{a}_{f0b}u^{(1)}_{e\tau} = e^{B^{(1)}}(\bar{q}^n_{0b}/\gamma\bar{p}^2_0)\{p^{(1)}_e - \bar{p}_0\rho^{(1)}_e\},$$

(8.25) $$q^{(1)}_{e\tau} + \bar{a}_{f0b}(\xi_\beta)^{-1}q^{(1)}_{e\beta} = -e^{B^{(1)}}(\bar{q}^n_{0b}/\gamma\bar{Q}\bar{p}^2_0)\{p^{(1)}_e - \bar{p}_0\rho^{(1)}_e\},$$

where $\bar{a}_{f0b}$ is the value of the ambient frozen sound speed at time T_b, when $\bar{q}_0$ has the value $\bar{q}_{0b}$. In view of (8.3), (8.8), (8.9), $1/\gamma\bar{p}^2_0$ must be replaced by $\gamma(1-b)^2$.

It is more revealing of the physics of this new situation to revert from the (τ, β)-coordinate system to a $(\tau, \hat{x})$-set of independent variables, where

(8.26) $$\frac{d\bar{x}}{d\hat{x}} = \sigma.$$

It is easy to make the transformation from one coordinate set to the other by making use of the definition of ξ in (5.3), of β via (5.4) and of (8.13) and (8.26) in the present section, with the result that (8.22)–(8.25) become, respectively

(8.27) $$\rho^{(1)}_{e\tau} + u^{(1)}_{e\hat{x}} = 0,$$

(8.28) $$u^{(1)}_{e\tau} + p^{(1)}_{e\hat{x}} = 0,$$

(8.29) $$p^{(1)}_{e\tau} + \bar{a}^2_{f0b}u^{(1)}_{e\hat{x}} = \gamma\bar{q}^n_{0b}(1-b)^2 e^{B^{(1)}}\left\{p^{(1)}_e - \frac{1}{\gamma}\bar{a}^2_{f0b}\rho^{(1)}_e\right\},$$

(8.30) $$\bar{Q}q^{(1)}_{e\tau} = -\gamma\bar{q}^n_{0b}(1-b)^2 e^{B^{(1)}}\left\{p^{(1)}_e - \frac{1}{\gamma}\bar{a}^2_{f0b}\rho^{(1)}_e\right\}.$$

Clearly (8.27) and (8.28) are basic small perturbation forms of the exact continuity and inviscid momentum equations (4.2), (4.3) in the new coordinate system. Although one can make a similar remark about (8.29), (8.30) vis-à-vis (4.4±), (4.7), with their transport terms omitted, it is evidently necessary to be just a little more careful to remember the strong influence of developments in the ambient atmosphere, as exemplified by the coefficient $\exp(B^{(1)})$, etc, on the right-hand sides of these equations.

Disturbances in pressure, density and velocity are all of the same amplitude, as with the uni-directional waves, but now the compositional perturbations $\bar{q}$ are also of this same order. Despite this fact the $\bar{p}, \bar{\rho}, \bar{u}$ problem is still decoupled from the $\bar{q}$ problem.

Most significantly, the equations themselves now betray no preference for propagation along $\bar{x}$-positive, *unlike* (7.23) which allows propagation in only this direction. Equations (8.27), (8.29) can be combined to give

(8.31) $$p^{(1)}_{e\hat{x}\tau} - \bar{a}^2_{f0b} p^{(1)}_{e\tau} = \Omega(\tau) \left\{ p^{(1)}_e - \frac{1}{\gamma} \bar{a}^2_{f0b} p^{(1)}_e \right\}$$

where

(8.32) $$\Omega(\tau) = \gamma \bar{q}''_{0b}(1-b)^2 \exp[B^{(1)}(\tau)].$$

Differentiating (8.31) with respect to $\hat{x}$, and taking the rather unusual step of defining a potential φ such that

(8.33) $$p^{(1)}_e = \varphi_{\hat{x}}, \quad u^{(1)}_e = -\varphi_\tau,$$

it can be shown that φ satisfies the equation

(8.34) $$\frac{\partial}{\partial \tau} \{\varphi_{\tau\tau} - \bar{a}^2_{f0b}\varphi_{\hat{x}\hat{x}}\} - \Omega(\tau) \left\{ \varphi_{\tau\tau} - \frac{1}{\gamma} \bar{a}^2_{f0b}\varphi_{\hat{x}\hat{x}} \right\} = 0.$$

It must be observed from (8.32) that the "relaxation frequency" Ω is irrecoverably time-dependent, since (8.14) shows that

(8.35) $$\Omega = \gamma \bar{q}''_{0b}(1-b)^2(1-\hat{b}\tau)^{-1}$$

and all variables in this relation are of order unity.

Equation (8.34) describes the progress of a "hierarchy" of waves with wave speeds between the "fast" adiabatic frozen sound speed $\bar{a}_{f0b}$ and the "slow" isothermal frozen sound speed $\bar{a}_{f0b}/\gamma^{1/2}$.

In view of the solution (8.14) for $B^{(1)}$, Ω is positive for all admissible τ values and the wave system must be unstable. In other words the amplitude of φ and hence, in general, of $u^{(1)}_e$, $\rho^{(1)}_e$, $p^{(1)}_e$ and $q^{(1)}_e$ will increase with time. The strong temporal variation of Ω makes it impossible to solve (8.34) by any of the various standard methods that are available for the constant coefficient type of equation.

When $\tau \to -\infty$, $\Omega \to 0$ and (8.34) goes over to the elementary classical wave equation for $u^{(1)}_e = -\varphi_\tau$ (taken at fixed $\hat{x}$, remember). The matching condition (8.18) is satisfied by a solution of this equation that represents an acoustic wave propagating along $\hat{x}$-positive; that is to say φ_τ behaves like a function of $\hat{x} - \bar{a}_{f0b}\tau$ in a neighbourhood of $\tau \to -\infty$. It is now clear that a nonzero value of the right-hand side of (8.34) begins to develop as time progresses, and that this begins to act as a source term for waves propagating in *both* directions, $\hat{x} \gtrless 0$. From the character of (8.34) it is clear that the wave system is dispersive and, as already mentioned above, the opposite of dissipative. There does not seem to be any alternative at this stage to a numerical solution of (8.34) but it is worth noting that, for a given value of γ, the details of the waves that are being created by interaction of the incident pulse with the ambient explosion are strongly dependent on the amplitude of the incident wave. This comes about through the links between σ, ϵ and b, as defined in (8.9) and, especially, via the swiftly changing relaxation frequency Ω, which (8.32) shows to be quite strongly dependent on the value of b.

As $\tau \to 1/\hat{b}$ the approximation that leads to (8.14) begins to break down, with an evident need to consider even shorter time scales than the one implied in (8.3). The general processes that will carry the solution forward into times T greater than $T_b + \sigma/\hat{b}$ are fairly clear, even if the details are not. The latter must await the outcome of any solution of (8.34) under condition (8.18). At some stage the consumption of reactant material must begin to play a direct role, in contrast to the decoupled part played by this quantity so far.

Derivation of a wave equation for the processes that arise as a result of uni-directional propagation breakdown of the type described in §7.3(vii a) is left as an exercise for the reader, who should take note of the comments about the scale of the $\partial/\partial\beta$ operation made in that section.

9. Nonlinear chemistry. When E, defined in §6.2, is of order unity the character of the equations in §6.4 changes. The exponential factor on the right-hand side of (6.17), (6.18) is now

$$\exp\{E(\gamma - 1)\bar{a}_{f0}^{-3}u^{(1)}\};$$

all of the quantities in the index of the exponential are of order unity, by hypothesis, and no further reductions are available through the use of parameter limits to simplify this function. From the opening remarks made in §7 it is clear that under present conditions the chemical effects are of *nonlinear* character; hence the title of this section.

It must be remarked that the disturbance-amplitude is *elevated* to the point that makes σ comparable with ϵ. The latter is a property of the combustible mixture, and is therefore not available for free selection; one should not think of diminishing ϵ to accord with some chosen small value of σ. The inference is that σ may therefore *not* be small, and certainly one could conceive of situations in which ϵ may be too large to make it and σ both comparable and small. However, one of the most significant features of gaseous combustion systems is their tremendous sensitivity to changes of temperature and this is exemplified with combustion-chemistry models like the one used here by relatively small values of ϵ. Table 1 gives some rough and ready ideas about the magnitudes of the quantities for a typical first order ($n = 1$) reaction, although it should be remarked that some quite wide variations are possible on this particular theme. The notion that ϵ is small is therefore generally acceptable, and further progress will be made by means of the large activation energy limit, $\epsilon \to 0$, $E = O(1)$.

FINITE AMPLITUDE WAVES IN COMBUSTIBLE GASES

9.1. The ambient atmosphere when $\epsilon \to 0$. The fact that $\epsilon \to 0$ has some useful consequences as far as ambient atmosphere behaviour is concerned. In particular an asymptotic solution for $\bar{p}_0$ given by

(9.1)
$$\frac{1}{\gamma \bar{p}_0} \sim 1 + \epsilon \ln(\gamma \hat{T} - \gamma T) \cdots ,$$
$$\gamma \hat{T} \sim 1 + \epsilon[2 + n(\gamma \bar{Q})^{-1}] \cdots ,$$

is good for all times $T < \hat{T}$ to within an exponentially (i.e. $\exp(-1/\epsilon)$) small error, and this fact will serve to simplify the forms of (6.17) and (6.18), in particular, in these time intervals. It should be observed that the ambient frozen sound speed $\bar{a}_{f0}$ behaves like

(9.2)
$$\bar{a}_{f0} = (\gamma \bar{p}_0)^{1/2} \sim 1 - \tfrac{1}{2}\epsilon \ln(\gamma \hat{T} - \gamma T) \cdots$$

and that, when (9.1) provides a valid estimate of $\bar{p}_0$, $\bar{q}_0$ remains within $O(\epsilon)$ of unity.

9.2. Perturbation equations. When disturbance amplitudes σ are comparable with the activation energy number ϵ, the equations for uni-directional wave propagation derived in §6.4 reduce to the following forms (note that the limiting process is now $\sigma \to 0$, with β, T fixed and $E = O(1)$; also use is made of the results in §9.1).

Equations (6.15), (6.16) become

(9.3, 4) $$p^{(1)} = \rho^{(1)} = u^{(1)},$$

while (6.17), (6.18) give

(9.5) $$2Eu_T^{(1)} = (\gamma \hat{T} - \gamma T)^{-1}\{\exp[(\gamma - 1)Eu^{(1)}] - 1\},$$

(9.6) $$\bar{Q}Eq_\beta^{(1)} = -\xi_\beta(\gamma \hat{T} - \gamma T)^{-1}\{\exp[(\gamma - 1)Eu^{(1)}] - 1\},$$

provided that $\sigma^2 \text{Re} = \epsilon^2 E^2 \text{Re} \to \infty$. In view of the fact that σ in the present chapter is essentially larger than the σ of §7.3 this proposition is already validated. The discussion in §§7.3(ii), (iii), (iv) also applies to the present situation, and results from these sections will be used here together with the simplifications that §9.1 affords for times $T < \hat{T}$.

It is worthwhile pointing out that (9.5) can be written with the diffusive term

$$\Delta(T)(\xi_\beta)^{-1}[(\xi_\beta)^{-1}u_\beta^{(1)}]_\beta = \Delta(T)u_{\Xi\Xi}^{(1)}$$

on the right-hand side, where $\Delta(T)$ and Ξ are defined in §7.1, if need be. The Burgers equation character of (9.5) is then apparent in the (Ξ, T)-coordinate system, just as it was for linear chemistry in (7.4). The only difference is that here the exponential term replaces the term that is linear in

$u^{(1)}$ in §7.1. It follows that any numerical solutions of (9.5) that involve numerical or artificial viscosity will actually be providing solutions of a Burgers-like equation with an augmented diffusivity $\Delta(T)$. On the whole this will only thicken any shocks that may appear in the system, but care must be exercised not to thicken them too much, especially in view of some matters that will be brought to light shortly in connection with local ignition.

9.3. General solutions. The general solution of (9.5) is easily written down; using the boundary condition from (5.12) then shows that

(9.7)
$$(\gamma - 1)Eu^{(1)}(\beta, T) = -\ln\left\{1 - [1 - e^{-(\gamma-1)ED'(\beta)}]\left(\frac{\hat{T} - \sigma\beta}{\hat{T} - T}\right)^m\right\},$$
$$m = (\gamma - 1)/2\gamma.$$

Equations (9.3), (9.4) together with (9.1), (9.2) can be used to show that (6.3) gives

$$\bar{a}_f \sim \bar{a}_{f0} + \tfrac{1}{2}\sigma(\gamma - 1)u^{(1)}(\beta, T),$$

or

(9.8) $$\bar{a}_f \sim 1 - \tfrac{1}{2}\epsilon \ln(\gamma\hat{T} - \gamma T) + \tfrac{1}{2}\epsilon(\gamma - 1)Eu^{(1)}(\beta, T)$$

in the present situation. Then (9.8) and (5.13) combine to produce the result

(9.9)
$$\xi - \beta + \epsilon ED(\beta) = \frac{1}{2E}\int_{\epsilon E\beta}^{T} \ln(\gamma\hat{T} - \gamma\tilde{T})\, d\tilde{T}$$
$$+ \cdots - \tfrac{1}{2}(\gamma + 1)\int_{\epsilon E\beta}^{T} u^{(1)}(\beta, \tilde{T})\, d\tilde{T}.$$

Making some elementary substitutions in the integrals (9.9) gives

(9.10)
$$\hat{\xi} = \xi + \frac{1}{2\gamma E}\int_{\gamma\hat{T}-\epsilon E\beta}^{\gamma\hat{T}-\gamma T} \ln s\, ds$$
$$= \hat{F}(\beta, T)$$
$$= \beta - \epsilon ED(\beta) + \frac{(\gamma + 1)}{E(\gamma - 1)^2}[\gamma\hat{T} - \epsilon E\gamma\beta]$$
$$\times \int_1^{[(\hat{T}-T)/(\hat{T}-\epsilon E\beta)]^{-m}} \ln(1 - \{1 - \exp[-(\gamma - 1)ED'(\beta)]\}\, s)s^{-1-1/m}\, ds.$$

Equations (9.9) and (9.10) give

$$-(\xi_\beta)^{-1}\bar{Q}q_\beta^{(1)} = \sigma\bar{Q}q_{\tilde{x}}^{(1)} = 2u_T^{(1)},$$

where $u_T^{(1)}$ is a time derivative taken with β fixed both here and in (9.11) below. By following arguments exactly like the ones used in §7.3(vii), it follows that

(9.11) $$\bar{Q}\sigma q^{(1)} = -2\int_{\tilde{x}}^{\tilde{x}_h(T)} u_T^{(1)}\, \partial\tilde{x}, \quad \tilde{x} < \tilde{x}_h(T),$$

where $\bar{x}_h(T)$ is the time-dependent location of the wave head. Of course $u^{(1)}$ is known, from (9.7), as a function of (β, T) and not of $(\bar{x}, T)$. Thus (9.11) does not provide a final evaluation of $q^{(1)}$ in terms of the "input" $D'(\beta)$. Despite this it will still prove to be quite a useful general result to have in the ensuing analysis.

Note that $u_T^{(1)}$ in (9.11) is found directly as a function of β and T from (9.5) and (9.7); (9.10) relates $\bar{x}$, T and β (since $\sigma\xi = T - \bar{x}$) and hence β and $\bar{x}$ for any given value of T.

9.4. Short pulse disturbances. If the piston input to the system is of limited duration, so that

$$D'(\beta \geqslant \beta_t) \equiv 0, \qquad \beta_t = O(1),$$

and $D(\beta)$ itself is bounded for all $\beta > \beta_t$ it follows that $\epsilon E\beta$ can be neglected relative to $\hat{T}$ in (9.10) while $\epsilon ED(\beta)$ itself can be neglected relative to β itself. Thus (9.10) goes over to

$$\hat{\xi} = \xi + \frac{1}{2\gamma E} \int_{\gamma\hat{T}}^{\gamma\hat{T} - \gamma T} \ln s \, ds = F(\beta, T) \tag{9.12}$$

to within $O(\sigma) = O(\epsilon)$, where

$$F(\beta, T) = \beta + \frac{(\gamma + 1)}{E(\gamma - 1)^2} \tag{9.13}$$
$$\times \int_1^{(1 - T/\hat{T})^{-m}} \ln(1 - \{1 - \exp[-(\gamma - 1)ED'(\beta)]\}s) s^{-1 - 1/m} \, ds.$$

A similar simplification occurs in (9.7), where the term in parentheses raised to the power m becomes $(1 - T/\hat{T})^{-m}$.

If no shock wave forms in the pulse, the wave head is at $\beta = 0$; with $D'(0) = 0$ it follows from (9.13) that $F(0, T) = 0$ and thence from (9.12), and the fact that $\sigma\xi = T - \bar{x}$, that

$$\bar{x}_h(T) - \bar{x}_t(T) = \sigma F(\beta_t, T).$$

Equation (9.13) makes it clear that $F(\beta_t, T)$ is $O(1)$ when β_t is $O(1)$, and it can now be seen from (9.11) that $q^{(1)}$ in a short pulse will remain $O(1)$ *provided* that $(u_T^{(1)})_\beta$ retains this character.

9.4(i). Local explosion. It is clear from the form of the solution in (3.1) that $u^{(1)}$ becomes logarithmically infinite on any characteristic, or constant-β line, when the time T is such that

$$(1 - T/\hat{T})^m = \{1 - \exp[-(\gamma - 1)ED'(\beta)]\}. \tag{9.14}$$

(The short pulse simplification that ignores $\sigma\beta$ relative to $\hat{T}$ is adopted here.)

If $D'(\beta) < 0$ the right-hand side of (9.14) is negative and there is no real value of T within the physically acceptable interval $0 \leq T < \hat{T}$ that satisfies the requirement. A negative $D'(\beta)$ value implies that $u^{(1)}$ itself is negative; with the present single direction of wave propagation, negative $u^{(1)}$ implies that the wave is one of expansion. When the wave is a compression $D'(\beta)$ and $u^{(1)}$ are positive; the right-hand side of (9.14) is positive and less than unity, so that a real time T_l in $0 \leq T_l < \hat{T}$ exists. In particular T_l is written to signify the *least* value of T that satisfies (9.14), and T_l will occur on the β-wavelet that carries the maximum value of $D'(\beta)$. Evidently this is the maximum amplitude of the compression wave.

The existence of an acceptable value of the time T_l in a compression wave is evidence of the onset of a localised ignition event, that is being propagated with the local wave speed through the ambient atmosphere. This feature of compression waves with amplitudes of order ϵ will be called "local explosion" from now on.

The ability to carry a mechanism for ignition out into the atmosphere, away from the source of the disturbance, and at local sound speeds, is clearly a physical process of considerable importance.

As remarked above, the earliest local ignition clearly occurs when $ED'(\beta)$ has its maximum positive value and it is interesting to list the values of this quantity that are associated with various reductions in T_l below the homogeneous value $\hat{T}$; this is done in Table 5 for $\gamma = 1.4$.

TABLE 5
Disturbance amplitude for various ratios of local-to-ambient explosion times $T_l/\hat{T}$. The velocity amplitude is measured as a fraction $\epsilon ED'(\beta)$ of the initial ambient sound speed a_{f0i}, and ϵ is $\frac{1}{30}$ here.

$\dfrac{T_l}{\hat{T}}$	$\dfrac{1}{2}$	$\dfrac{2}{3}$	$\dfrac{3}{4}$	$\dfrac{4}{5}$	$\dfrac{5}{6}$	$\dfrac{6}{7}$	$\dfrac{7}{8}$	$\dfrac{8}{9}$
$ED'(\beta)$	5.9	4.8	4.3	4	3.7	3.5	3.4	3.3

9.4(ii). *Comments on disturbance magnitude.* The boundary condition (5.12) shows that

$$\bar{u} = \sigma D'(\beta) = \epsilon ED'(\beta)$$

at the piston face, so that $ED'(\beta)$ is proportional to the amplitude of the dimensionless input velocity, u/a_{f0i}. Since ϵ is likely to be in the range $\frac{1}{20}$ to $\frac{1}{40}$, as can be seen from Table 1, it is first of all clear that the notion of a small perturbation is being pushed rather close to its limits of validity if, in present circumstances, one is modelling an event that leads to local explosion times significantly smaller than the ambient homogeneous value $\hat{T}$. It is reasonable to interpret the word "significantly" here to mean smaller than $\hat{T}[1 - O(\epsilon)]$; for instance, if $T_l < \hat{T}(1 - \epsilon) \leq 0.95\hat{T}$ it is correct to think of the local explosive event as having taken place in an atmosphere whose pressure has varied from its initial state by much less than 10–15%, as can be seen from (9.1).

Certainly small perturbations as large as the ones envisaged in the foregoing paragraph have been modelled in other branches of aerodynamics by methods that are, in principle, exactly like the ones being used here. It must be remembered that the present perturbation scheme is *not* a linearising process. Since one can anticipate a good measure of qualitative exactitude from the methods employed here, it is worth proceeding for that end alone; experience suggests that quantitative accuracy is achievable by the use of suitable second approximations, but they have not been pursued up to this point in the research.

Apart from anything else it is valuable to be able to use the short pulse criterion (9.14), and its numerical illustration in Table 5, as an indication of the magnitude of those gas dynamic disturbances that are capable of advancing the ignition time, locally, to an extent that is significant enough to involve some new physical phenomena when compared with the previous smaller disturbance case, discussed in §§7 and 8.

9.4(iii). *Local explosion in oscillatory waves.* The general local explosion criterion, that is the criterion for disturbances that are *not* of the short pulse kind, follows from (9.7), namely

$$(9.15) \quad (1 - T_I/\hat{T}) = (1 - \sigma\beta/\hat{T})\{1 - \exp[-(\gamma - 1)ED'(\beta)]\}^{1/m}.$$

If the signal launched by the piston is harmonic, for example, so that $D'(\beta)$ goes through a succession of positive maxima all of the same magnitude, (9.15) shows that it will be at the *first* of these maxima that the local explosive processes will be initiated, since the factor $(1 - \sigma\beta/\hat{T})$ diminishes monotonically as β grows from zero. It is not a priori obvious that this will be so as successive maximum compressions are sent into an atmosphere whose chemical activity is growing. It must be noted that one cannot use the general criterion given here for any disturbance-inputs that violate the basic requirement for $q^{(1)}$ to be $O(1)$. (9.11) shows that $q^{(1)}$ will only be $O(1)$ when $\bar{x}_h(T) - \bar{x}$ is $O(1)$ if the sign of $u_T^{(1)}$ alternates suitably; hence the present limitation to oscillatory waves.

9.4(iv). *Shock waves and local explosion.* The formation of shocks within a compression wave is a matter that is unaffected in general by the increased $O(\epsilon)$ amplitudes that are being considered in this chapter; it is only the details of the shock fitting process that are made more awkward by the present need to work with solutions like (9.7) and (9.10). To reiterate, the general forms of these solutions do not differ from those found in §7, in particular (7.24) and (7.26) or (7.32), and the discussions and analyses in §§7.3(ii), (iii), and (iv) come through unchanged to the present case, provided that one remembers that while $\bar{x}$ in (7.32) is a general result, (7.35) *is not*. This is because $u^{(1)}$ is not simply a product of a function of β and function of T in the present case (see (9.7)) as it is in §7 (see (7.24)). Thus while (7.37) is true in general circumstances, (7.38) is not.

Details of the way in which (7.37) is exploited to fit a shock wave into, say, a short compression pulse of sinusoidal form

$$(9.16) \qquad D'(\beta) = \begin{cases} a \sin b\beta, & 0 \leq b\beta \leq \pi, \\ 0, & b\beta < 0, \quad \pi < b\beta, \end{cases}$$

with $a \neq 0$ and $b > 0$, are too lengthy to display here. The results of the process are interesting and will be exemplified by some figures taken from a paper by the writer. In brief this work shows that;

a) the shock wave forms just behind the leading wavelet $\beta = 0$ (cf. §7.3(v) for the case $\sigma \ll \epsilon$);

b) the shock wave may grow beyond the original peak of the disturbance, so that the local explosion event coincides with the shock wave (see Figs. 1, 2 and also note the remarks in §9.2 about shock thickness);

c) local explosion may occur some distance behind the shock (Fig. 3);

d) local explosion may occur before a shock wave forms in the continuous compression pulse.

It is interesting to see how, once formed, a shock's strength grows with time, and Fig. 4 illustrates three situations and compares them with the development of the shock that appears at the head of the disturbance described in (9.16) in a chemically inert atmosphere.

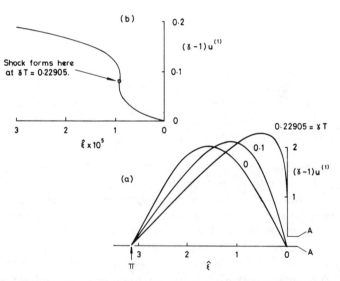

FIG. 1. *The progress of the short sinusoidal pulse of compression (see* (9.16)), *whose profile is* $2 \sin \hat{\xi}$ *at time* $T = 0$. (a) *Note the growth in the amplitude of the disturbance above the value 2 as time proceeds up to the instant of shock formation.* (b) *Details of the profile at the instant of shock formation just downstream of the wave head.*

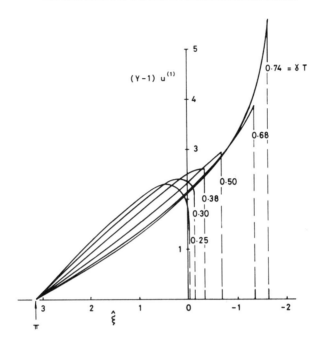

FIG. 2. *Continuation of the profiles of Fig.* 1 *into times after shock formation. The rapid growth of shock wave amplitude is illustrated by the size of the discontinuity that travels into $\hat{\xi} > 0$ as T increases. The shock wave "swallows" the smooth peak at $\gamma T \simeq 0.38$ and thereafter the approach to a logarithmically infinite value of disturbance (local explosion) occurs immediately downstream of the shock as $\gamma T \to 1$.*

9.4(v). *Behaviour near an isolated local explosion.* As explained in (iv) above, the local ignition event may take place at the compression peak in isolation from other features such as the appearance of a shock wave. Analysis of the short pulse forms of the solutions in §9.3 then shows that in the neighbourhood of the maximum value of $D'(\beta)$ at $\beta = \beta_m$, say, $u_T^{(1)}$ increases like $1/\Delta T$, where

(9.17) $$\Delta T = (T_l - T)/\hat{T},$$

while $u_\beta^{(1)}$ is proportional to $1/\sqrt{\Delta T}$ for β in the interval

(9.18) $$\Delta\beta = \beta - \beta_m \propto \sqrt{\Delta T}.$$

Thus as ΔT becomes small and of order $\delta(\epsilon) = o(1)$ as $\epsilon \to 0$, for example, the solutions in §9.3 begin to break down and revised approximations for $\bar{u}$, $\bar{p}$, $\bar{\rho}$ and $\bar{q}$ all become necessary.

Note that (9.11) once again signals that the need for these revisions coincides with the fact that $\bar{q}$ begins to exceed $O(\epsilon^2)$.

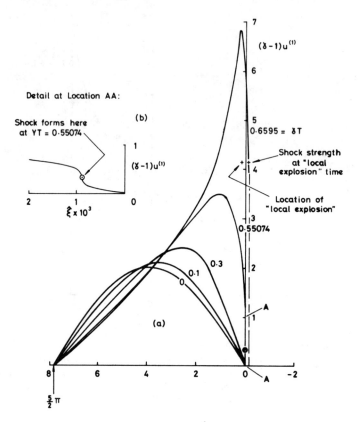

FIG. 3. (a) *The pulse shape is now initially* $2\sin(\tfrac{2}{5}\hat{\xi})$. *Local explosion occurs at the smooth pulse peak shortly after* $\gamma T = 0.63595$ *and downstream of the shock (see points marked* +). (b) *Details at the instant of shock formation.*

Analysis quite similar to that described in §8.3 (albeit for failure of unidirectional wave propagation arising from a different source; see §§8.1, 8.2) shows that the new wave field that emerges in the neighbourhood of local explosion obeys the following equations, to leading order.

First define τ and $\hat{x}$ via

(9.19) $\qquad T = T_l + \tau\delta, \quad \bar{x} = \bar{x}_l + \hat{x}\delta, \quad \delta = (\epsilon E)^2,$

where $\bar{x}_l$ is $\bar{x}$ at β_m and T_l. Then, with

$$\psi \sim -\frac{\epsilon}{(\gamma-1)}\ln\delta + \epsilon\psi_e^{(1)}(\tau,\hat{x}),$$

(9.20) $\qquad \psi = p, \rho, u,$

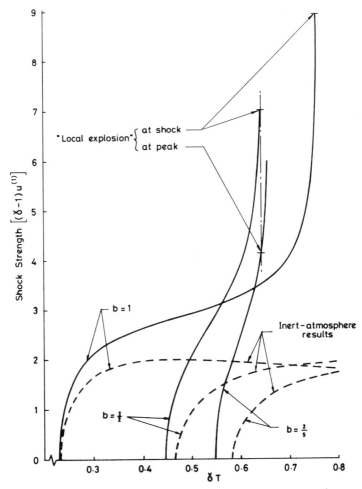

FIG. 4. *The growth of shock strength with time for the initial profile* $2 \sin b\hat{\xi}$ *for three values of b, compared with behaviour from the same initial state in a chemically inert atmosphere. The exceptional rate of growth of shock strength for the longer pulses is noteworthy.*

it is found that

(9.21) $\quad \rho^{(1)}_{e\tau} + u^{(1)}_{e\hat{x}} = 0,$

(9.22) $\quad u^{(1)}_{e\tau} + p^{(1)}_{e\hat{x}} = 0,$

(9.23) $\quad p^{(1)}_{e\tau} - \rho^{(1)}_{e\tau} = (\gamma\hat{T} - \gamma\hat{T}_l)^{-1} \exp\{\gamma p^{(1)}_e - \rho^{(1)}_e\}.$

The $\bar{q}$-problem remains decoupled from the $\bar{p}, \bar{\rho}, \bar{u}$ field at this stage, but variations in $\bar{q}$ are $O(\epsilon)$, as are the variations in $\bar{p}$, etc.

Comparison of (9.21)–(9.23) with (8.27)–(8.29) reveals very close, superficial, similarities. The most obvious difference is on the right-hand side of (9.23), which consists of a constant factor times the exponential that signifies the response of chemical reaction rates to the existence of disturbances; (8.29) makes the latter linear in the disturbance quantities, but has a function of time for their coefficient. However $\partial/\partial\tau$ and $\partial/\partial\hat{x}$ represent rates of change that are of order ϵ^{-2} here, relative to order one changes in the ambient atmosphere over an induction time interval, whereas they are of order σ^{-1} in §8.3.

The appearance of sets of equations like (9.21)–(9.23) is a recurrent feature of wave propagation in the presence of imminent ignition, and the associated elevation of reactant consumption to a leading order status along with the gas dynamical disturbances. The implied simultaneous production of waves that travel in either direction is clearly associated with the strictly scalar character of local density and pressure changes arising from deposition of combustion energy into the wave field.

9.5. A sustained disturbance. As a contrast to the short-pulse disturbances described in §9.4 it is instructive to examine the behaviour of a disturbance that is sustained over a period of time. A simple example has already been mentioned in §7.3(vii a) and this is reiterated here, namely the step function

$$(9.24) \qquad D'(\beta) = \begin{cases} 0, & \beta < 0, \\ 1, & \beta > 0. \end{cases}$$

This will clearly lead to the immediate establishment of a (weak) shock wave at the wave head whose subsequent progress must follow the path laid down for it by (7.37).

9.5(i). *Shock fitting.* It can be seen that $\bar{x}$ in (7.37) can be replaced by the quantity $\hat{\xi}$ defined in (9.10). Making use of the result that $\hat{\xi}$ is equal to $\hat{F}(\beta, T)$ from that same equation the shock fitting rule (7.37b) can be translated into

$$(9.25) \qquad \dot{\beta}_+ \frac{\partial \hat{F}(\beta_+, T)}{\partial \beta_+} + \dot{\beta}_- \frac{\partial \hat{F}(\beta_-, T)}{\partial \beta_-} = 0$$

where $\dot{\beta}_\pm$ is the time derivative of $\beta_\pm$ where these characteristics meet on the shock wave.

(9.10) shows that $\hat{F}(\beta_-, T)$ is simply equal to β_- for all $\beta_- < 0$, as required by (9.24). Thus the last term in (9.25) becomes simply $\dot{\beta}_-$. Since on the shock wave $\hat{\xi}$ must have the same value, for a given T, for β equal to either β_+ or β_-, i.e.

$$(9.26) \qquad \hat{\xi}_s = \hat{F}(\beta_+, T) = \hat{F}(\beta_-, T) = \beta_-$$

when (9.24) holds, β_- can be eliminated between (9.25) and (9.26). The result is a relation, a first order nonlinear differential equation, for β_+ as a function of T, namely

(9.27) $$\frac{d\beta_+}{dT} = -\tfrac{1}{2}\frac{[\partial \hat{F}(\beta_+, T)/\partial T]_{\beta_+}}{[\partial \hat{F}(\beta_+, T)/\partial \beta_+]_T}.$$

The reader will quickly verify that even the simple function (9.24) when incorporated into (9.7) and (9.10), does not make for simple analytical solution of (9.27); the shock fitting problem with nonlinear chemistry is essentially a numerical exercise.

However, the weakness of the waves defined by (9.24) allows one to infer that the shock speed will not greatly exceed the value $\bar{a}_{f0}$ and that the wave head $\bar{x}_h(T)$ is in some near vicinity of the path $\bar{x}_h \approx T$. It follows at once from (9.11) that the perturbation $\bar{q}$, namely $\sigma^2 q^{(1)}$, will begin to exceed $O(\sigma^2)$ in the neighbourhood of the piston face ($\bar{x} = \sigma T$ in the present case; see (5.1)) for any T-value that is itself in excess of $O(\sigma)$. A breakdown of the present theory, of the type described in §7.3(vii a), will therefore occur after an interval of time of order σ.

Note that local explosion of the type described in §§9.4(i), (iii) is not involved in the breakdown of the theory just described; since T_l is essentially $O(1)$, as can be seen from (9.15) for example, one can deduce that modification of the present theory is required before local explosion, of the type described so far, has had a chance to occur.

9.5(ii). *Revised perturbation equations.* It is evidently necessary to revise the order of magnitude of $\bar{q}$ to read

$$\bar{q} \sim \sigma q^{(1)}$$

at some early time in a sustained disturbance. However $\bar{u}$, $\bar{p}$ and $\bar{\rho}$ will all remain of basic order σ, since there is no sign of impending failure in their asymptotic parameter-perturbation representations. Neither is there any need to revise the time and space derivative orders that are implicit in (4.2), (4.3), (4.4) and (4.7).

Thus with $\sigma = \epsilon E$, using the limit $\epsilon \to 0$ with $\bar{x}$, T fixed, and writing

(9.28) $\quad\quad \psi \sim \epsilon \psi^{(1)}(T, \bar{x}), \quad \psi = p, \rho, u, q$

the equations from §4.1 give

(9.29) $\quad \rho_T^{(1)} + u_{\bar{x}}^{(1)} = 0,$

(9.30) $\quad u_T^{(1)} + p_{\bar{x}}^{(1)} = 0,$

(9.31) $\quad p_T^{(1)} + u_{\bar{x}}^{(1)} = (\gamma \hat{T} - \gamma T)^{-1}\{\exp[\gamma p^{(1)} - \rho^{(1)}] - 1\},$

(9.32) $\quad \bar{Q} q_T^{(1)} = -(\gamma \hat{T} - \gamma T)^{-1}\{\exp[\gamma p^{(1)} - \rho^{(1)}] - 1\}.$

Comparison of (9.29)–(9.31) here with (9.21)–(9.23) and (8.27)–(8.29) makes it clear that there is a general similarity of physical behaviour in all these situations, a view that is reinforced by the behaviour of $\bar{q}$ in each case. It is equally important to take note of the *differences,* particularly of time and length scales, which are not the same in any of the examples.

9.5(iii). *Perturbation equation for the temperature.* It is helpful to reiterate (4.5b) here, namely

$$\bar{\theta} = \gamma(\bar{p}_0 + \bar{p})(1 + \bar{\rho})^{-1},$$

since it is then clear that definition of $\theta^{(1)}$ via

(9.33) $$\bar{\theta} = 1 + \epsilon\theta^{(1)}$$

leads to the relation

(9.34) $$\theta^{(1)} = -\ln(\gamma\hat{T} - \gamma T) + \gamma p^{(1)} - \rho^{(1)},$$

where (9.1) has been used to eliminate $\bar{p}_0$.

It is essential to remember that $\theta^{(1)}$ as defined in (9.33) is the *complete* perturbation of temperature, ambient atmosphere plus disturbance, measured from the ambient initial value of unity.

Equation (9.34) can now be used in its turn to eliminate $p^{(1)}$ from the four small-disturbance conservation equations. The first three of these become

(9.35) $$\rho^{(1)}_T + u^{(1)}_x = 0,$$

(9.36) $$\gamma u^{(1)}_T + \rho^{(1)}_x + \theta^{(1)}_x = 0,$$

(9.37) $$\rho^{(1)}_T + \theta^{(1)}_T + \gamma u^{(1)}_x = \gamma \exp(\theta^{(1)}),$$

and it is now easy to eliminate $u^{(1)}$ and $\rho^{(1)}$ and so derive an equation for $\theta^{(1)}$ alone. It is helpful to write this equation in two ways, either

(9.38a) $$\frac{\partial^2}{\partial T^2}\{\theta^{(1)}_T - \gamma \exp(\theta^{(1)})\} - \frac{\partial^2}{\partial x^2}\{\theta^{(1)}_T - \exp(\theta^{(1)})\} = 0$$

or

(9.38b) $$\left\{\frac{\partial^2}{\partial T^2}(\theta^{(1)}_T) - \frac{\partial^2}{\partial x^2}(\theta^{(1)}_T)\right\} - \left\{\gamma\frac{\partial^2}{\partial T^2}(e^{\theta^{(1)}}) - \frac{\partial^2}{\partial x^2}(e^{\theta^{(1)}})\right\} = 0.$$

These two versions (9.38a) and (9.38b) of the same equation are both illuminating, and shed light on the physics of the situation that they represent.

Consider (9.37) for a moment; if $\rho^{(1)}$ is zero, then the density of the whole system, background *and* disturbance, is fixed and the remnant of (9.37), namely

(9.39) $$\theta^{(1)}_T - \gamma \exp(\theta^{(1)}) = 0,$$

represents the early-time or induction phase of an explosive event which takes place at constant volume. If we now assume that both $p^{(1)}$ *and* the perturbation in ambient pressure vanish, it follows from (9.34) that $\rho^{(1)}$ is simply $-\theta^{(1)}$; (9.35) and (9.37) show that

$$-(\gamma - 1)\rho_T^{(1)} + \theta_T^{(1)} = \gamma \exp(\theta^{(1)})$$

so that under conditions of constant pressure we find that

(9.40) $$\theta_T^{(1)} - \exp(\theta^{(1)}) = 0.$$

This simple result should be compared with (9.39); evidently (9.40) represents the early-time temperature behaviour under contrasting conditions of constant pressure.

The meaning of the operators in { } in (9.38a) is now clear; they represent, in order, the early-time effects of explosions at constant volume and explosion at constant pressure.

Equation (9.38b) displays the role of the remaining operators in the partial differential equation for $\theta^{(1)}$. The first term in { } is the classical wave-equation operator, operating on $\theta_T^{(1)}$. Since it contains the derivatives of highest (third) order, it will determine the characteristics of (9.38), which are clearly the lines $d\bar{x}/dT$ equal to either $+1$ or -1. From the nondimensionalisations of §4.1 these are evidently wavelets whose speed of propagation is equal to the initial unperturbed frozen isentropic sound speed (cf. (2.14)). The second pair of { } brackets in (9.38b) also contains an elementary wave operator but in this case the associated propagation speed is given by $|d\bar{x}/dT|$ equal to $1/\sqrt{\gamma}$. In view of (2.14) this represents a wave speed, in dimensional terms, given by $\sqrt{p/\rho}$ evaluated under ambient initial conditions. This is the isothermal or Newtonian sound speed calculated under chemically frozen conditions, and we are now in a position to describe the physical implications that lie behind (9.38).

With the appearance of an upper (isentropic) and a lower (isothermal) wave speed the system is dispersive, and this dispersive wave system is intimately involved in the decision about whether ignition should proceed under local conditions of constant pressure or constant volume (cf. (9.39) and (9.40)).

9.5(iv). *Unstable character of temperature equation solutions.* A last and significant feature of solutions of (9.38) is that they are unstable. That is to say, they will become unbounded in a finite time. The feature is directly associated with the minus sign between the operators in (9.38b) as will be demonstrated here, briefly.

Suppose that propagation is predominantly along $\bar{x}$-positive at unit speed. With the hypothesis that rates of change are slow *along* such a wavelet, compared with rates of change from one wavelet to another, it is roughly true that $\partial/\partial T \simeq -\partial/\partial \bar{x}$. Simplifying (9.38b) on this basis, integrating twice with respect to T and simply discarding the concomitant functions of $\bar{x}$ for present purposes, one finds that

$$\text{(9.41)} \qquad \theta_T^{(1)} + \theta_x^{(1)} = \left(\frac{\partial \theta^{(1)}}{\partial T}\right)_\xi \simeq \tfrac{1}{2}(\gamma - 1)e^{\theta^{(1)}},$$

where $\tilde{\xi} = T - \bar{x}$. If $\theta_i^{(1)}(\bar{x}, 0)$ is the initial value of $\theta^{(1)}$, (9.41) has a solution

$$\theta^{(1)} \simeq -\ln\{\exp[-\theta_i^{(1)}(-\tilde{\xi}, 0)] - \tfrac{1}{2}(\gamma - 1)T\}$$

and this grows with time until it becomes unbounded when $\tfrac{1}{2}(\gamma - 1)T$ is equal to $\exp[-\theta_i^{(1)}(-\tilde{\xi}, 0)]$.

The reader will find it instructive to follow a similar line of reasoning for wavelets that propagate with speed $1/\sqrt{\gamma}$, when he or she will find that $\theta^{(1)}$ obeys an "anti-diffusion" equation (diffusion equation with a negative diffusion coefficient) whose solutions have a self-focusing property that turns a mildly positive $\theta^{(1)}$-perturbation into a Dirac δ-function-like disturbance peak.

10. The effects of a large disturbance. The course taken by this account of wave propagation in a combustible atmosphere has been one marked by a steady rise in the amplitude of the input disturbances, from truly acoustical levels in §7.2, through weak waves with linear chemistry in §7.3, and on to not-so-weak waves and nonlinear chemistry in §9. It is therefore appropriate to end on what might be thought of as the top rung of the disturbance ladder, that is to say, with disturbances whose order of magnitude is unity. In fact it will be much more the "initiating" disturbance that is of substantial amplitude. What happens behind this wave head will still fall within the compass of the perturbation theory that has been the sole concern so far, but it will at least illustrate another facet of the utility of the small disturbance idea.

10.1. A strong shock wave. It will be assumed that a strong shock wave propagates into an otherwise unperturbed but exothermically reacting atmosphere, of the type described in §3. Since induction times (see (3.7)) are proportional to $\epsilon \exp(1/\epsilon)$ a modest reduction in ϵ will lead to very large increases in induction time, with the consequence that it is perfectly reasonable to think of the ambient atmosphere ahead of the shock (see Fig. 5) as effectively chemically inert. Behind the shock t_I has reduced to an extent that makes post-shock ignition inevitable within a short time, and therefore also distance, from initiation of the leading shock wave (cf. Table 1). One may think of the latter as having been produced by an impulsive piston movement or, more practically perhaps, by the reflection of a plane incident shock from the closed end of the tube in which it is propagating.

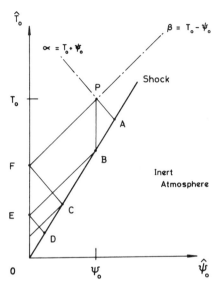

FIG. 5. *Wave field between a piston ($\Psi_0 = 0$) and a strong shock, ahead of which the atmosphere is effectively chemically inert. Ψ_0, T_0 are Lagrangian coordinates of a typical point P in the field $\hat{\Psi}_0$, $\hat{T}_0$ and α, β are characteristics through P. The line BP is the path of the element of gas that crossed the shock at point B; other letters help to identify particular wavelets (see Fig. 6). The initial induction time at point 0 behind the shock is t_I.*

With the assumption of inert conditions ahead of the shock the problem reduces to the need to solve for behaviour in the reacting gas between wall and shock (Fig. 5). It is not difficult to see, especially with the aid of results like those in §9.5, that one needs to treat the field in the manner implied by the representations (9.28), with ϵ given by the activation energy normalised by conditions immediately downstream of the strong initiating shock wave in its initial state. Referring to (3.6) this means that p_{0i} and ρ_i are conditions behind the shock as time $T \to 0+$. In the circumstances the post-shock field will be described by (9.38), whose solution must therefore be sought subject to appropriate velocity conditions at the piston or endwall *and* at the shock itself. The latter now constitutes a free boundary whose position must be calculated as part of the problem.

The solutions, whose predictions will be described briefly below, were also obtained for rather more general circumstances, in which the solid wall or piston is replaced by a gas/gas contact surface, and by a slightly different route. It is often advantageous to use not $\bar{x}$, T but Ψ, T where Ψ is the Lagrangian coordinate

$$\rho_i a_{f0i} t_I \Psi = \int_{\bar{x}_p(T)}^{\bar{x}} \bar{\rho}(\bar{x}, T) \, d\bar{x}$$

and $\bar{x}_p(T)$ is the piston or contact-surface path, in problems of one-dimensional unsteady motion of a compressible fluid. When this is done, the effect on (9.38) is formally minimal and simply replaces $\bar{x}$ by Ψ in the derivatives; the *form* of the equation is unaltered but interpretation of the wave motion is modified somewhat since an element of convection is implicit in the use of Ψ in place of $\bar{x}$ (formally similar parameter perturbations give rise to slightly different results in one coordinate system when compared with another).

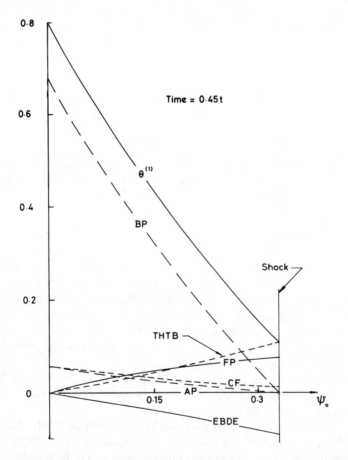

FIG. 6. *The temperature perturbation function $\theta^{(1)}$ (see §9.5) versus Ψ_0 for a time T_0 equal to $0.45t_1$. The letters identify the contributions to $\theta^{(1)}$ that come from the various wavelets and paths shown on Fig. 5; THTB refers to the temperature increment $\underline{at}$ the accelerating shock wave. The particle path BP provides a major contribution, as does $\overline{THTB}$ itself near the shock, but the effect of the disturbances transmitted along FP, AP, CF and EBDE (wavelets EB and DE combined) is not negligible.*

It is actually also advantageous to go back yet one more step and use the pair of exact characteristic parameters as the independent variables. This makes T and Ψ into dependent variables for which additional asymptotic series must be written down. To leading order the characteristic parameters α and β are simply related to Ψ_0 and T_0 (the leading order values of $\Psi = \Psi(\alpha, \beta)$ and $T = T(\alpha, \beta)$) as follows:

$$\Psi_0 - T_0 = -\beta, \qquad \Psi_0 + T_0 = \alpha.$$

The next-order terms, equal to $O(\epsilon)$, allow one to keep a check on the development of nonlinear wave processes and these are found to be insignificant right up to the point of local ignition at the wall, piston face or contact surface.

Solutions of (9.38) are carried out by a simple iterative numerical method *after* having converted the partial differential equation into an integral equation. Two results of some general interest are exhibited here.

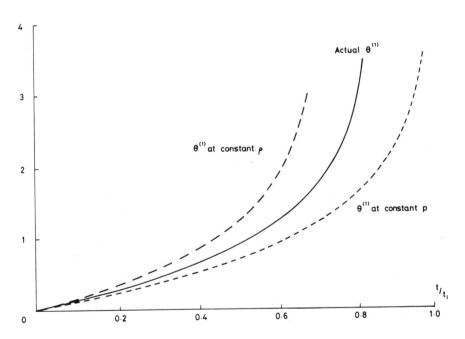

FIG. 7. *Temperature perturbation at the face of the constant speed piston versus time, compared with induction period behaviour at constant volume (or ρ) and constant pressure p. The latter two curves are $\ln(1 - t/\alpha t_I)$ where $\alpha = 1/\gamma$ or 1 for constant ρ or constant p respectively. The actual induction time is $\frac{1}{2}(1 + 1/\gamma)t_I$ to leading order accuracy.*

Figure 6 shows temperature versus "distance" (or Ψ_0) for a given time $T_0 = 0.47$, and displays the obviously important role of heating within the fluid element itself (BP in the figure, and note Fig. 5). The part played by the wave processes, although smaller than the internal heating of a fluid element, obviously cannot be ignored. The most dramatic testimony to the combination of physical effects described towards the end of §9.5(iii) is illustrated in Fig. 7. This shows that the actual ignition event, that takes place at the piston or contact surface in this case, does so at essentially the arithmetic mean of the constant-volume and constant-pressure times (namely $1/\gamma$ from (9.39), and 1 from (9.40)).

Thus not only does the dispersive wave system serve to share out the combustion physics between constant-volume and constant-pressure limits, but it does so without bias!—at least up to the onset of vigorous local reaction.

Concluding remarks. The essence of the work described here is to be found in the association of unsteady gas dynamical processes with energy liberating combustion reactions. Since the latter are invariably highly sensitive to changes of local temperature, the elevated values of temperature that travel with a gas dynamical wave of compression lead to local increases in the intensity of the combustion reaction; this source of energy, which is intrinsically coherent with the compression wave, results in an increase in the wave's amplitude and an unstable feedback process is established. Addition (subtraction) of energy in a region of high (low) pressure is generally referred to as Rayleigh's criterion for the amplification of disturbances; the energy source is simply the combustion reaction in the present case. That a naturally unstable state of affairs exists for all combinations of wave strength and intensity of exothermic chemical activity is fairly obvious. What is not so obvious is its extent and significance in the general context of gas phase phenomena. This essay provides at least some of the answers to these questions by identifying the capacity of combustible atmospheres to sustain acoustic waves against viscous damping, to amplify weak shock waves, and to propagate and advance ignition locally, just to nominate a few of the key results of the analysis. This all begins to establish a theoretical basis for some of the phenomena observed in the birth of detonation and quasi-detonation waves. The reader is encouraged to consult the article by Professor Lee that is listed in the References; that there is much to do before quantitative theories of these transient processes can be thought of as at all complete will then become very evident to him or her. Hopefully the present article will have convinced the reader that some rational lines of approach towards these ends do now exist.

List of symbols

a_f	frozen sound speed	W_m	mixture molecular weight
$A(T)$	amplification factor [defined in (7.5)]	x	distance
$\bar{B}, \bar{B}_m, B^{(1)}$	defined in §§6.5 and 8.1	$\bar{x}_h$	wave head position (dimensionless)
$B(\beta)$	defined in (7.36)	$\bar{x}_p$	piston position (dimensionless)
C_{pf}	frozen specific heat at constant pressure	$\bar{x}_t$	wave tail position (dimensionless)
D, D'	piston-displacement function and its derivative	$\hat{x}$	defined in (8.26) and (9.19)
E	defined in §6.2	$\mathcal{D}$	mass diffusion coefficient
E_A	activation energy	$\mathcal{P}$	pre-exponential frequency factor
$f(T)$	amplification factor (defined in (7.25))	$\mathcal{Q}$	energy-flux vector
$\hat{F}, F$	defined in (9.10) and (9.12)	$\mathcal{R}$	combustion reaction rate
g	diffusive mass flux	α	characteristic parameter (§10.1)
h	specific enthalpy	β	characteristic parameter
ln	natural logarithm	β_t	β at wave tail (§9.4)
M	amplitude parameter (§7.2)	β_m	defined in §9.4(v)
n	order of combustion reaction	γ	frozen specific heat ratio
p	pressure	δ	parameter defined in (9.19)
Pr	Prandtl number	δ_0	diffusivity of sound, (6.11)
q	reactant mass fraction	Δ	dimensionless diffusivity, (7.6)
$\hat{Q}$	combustion energy per unit mass of reactant	ϵ	activation-energy number, (3.6)
$\bar{Q}$	dimensionless combustion energy, (4.8b)	η	viscosity
		θ	temperature
R	universal gas constant	λ	thermal conductivity
Re	Reynolds number	ξ	wavelet coordinate, (5.3)
Sc	Schmidt number	$\hat{\xi}$	defined in (9.10)
t	time	Ξ	wavelet coordinate, (7.3)
t_I	induction time	ρ	density
T	re-scaled time	σ	disturbance amplitude parameter
$\hat{T}$	asymptotic estimate of T at induction time, (9.1)	σ_1	time-scale parameter (§5.1 and §6.2)
T_b	defined in (9.19)	τ	re-scaled dimensionless time, (8.13) and (9.19)
T_I	local ignition time (§9.4(i))	τ_v	viscous stress tensor
T_0	defined in §10.1	φ	potential, (8.33)
ΔT	time interval, (9.17)	Ψ	Lagrangian coordinate (§10.1)
u	gas velocity	Ψ_0	see §10.1
$\hat{u}_p$	piston velocity (dimensionless)	ω	frequency
W	reactant molecular weight	Ω	see (8.35)

Superscripts and subscripts

$(\ )_0$	ambient-atmosphere value
$(\ )_{0i}$	initial ambient atmosphere value
$(\ )_i$	initial value
$(^-)$	dimensionless value, see §4.1

REFERENCES

This list, which is not intended to be a complete inventory of works on waves in explosive atmospheres, contains items which have been consulted during the preparation of the present article. They should provide both a broad view of recent work on the present topic, and a means of access to the quite substantial literature on the subject of waves and reaction or relaxation effects.

[1] G. E. ABOUSIEFF AND T. Y. TOONG, *Nonlinear wave-kinetic interactions in irreversibly reacting media*, J. Fluid Mech., 103 (1981), pp. 1–22. This paper derives an equation like (6.17) and augments the diffusivity δ_0 ("artificial viscosity") so as to obtain numerical solutions for the development of initial N-waves; it supplements work in §9.4.

[2] ———, *Theory of unstable one-dimensional detonations*, Combustion and Flame, 45 (1982), pp. 67–92. This paper considers the one-dimensional unsteady very small amplitude (cf. §7) wave-field downstream of a plane steady detonation (comprised of shock, induction domain, rapid-reaction region and equilibrium zone). Several ad hoc approximations are made, together with the assumption $\epsilon \ll 1$. It shows the importance of wave/reaction coupling as the driving mechanism for the acoustical type of field; note that transport effects are not required in either ambient atmosphere structure (cf. the fast flame studies of Clarke) or the wave system.

[3] G. E. ABOUSIEFF, T. Y. TOONG, AND J. CONVERTI, *Acoustic and shock kinetic interactions in non-equilibrium H_2-Cl_2 reactions*, 17th Symposium (International) on Combustion, The Combustion Institute, Pittsburgh, 1979, pp. 1341 ff. An important paper that reports experimental investigations of a few wavelengths of piston input to an ambient atmosphere, undergoing a H_2-Cl_2 reaction, which validates results such as those in §7.2(i). Also refers to earlier papers by Toong and his co-workers.

[4] P. A. BLYTHE, *Wave propagation and ignition in a combustible mixture*, 17th Symposium (International) on Combustion, The Combustion Institute, Pittsburgh, 1979, pp. 909–916. The waves here are predominantly waves of expansion which may suppress ignition.

[5] J. F. CLARKE, *Diffusion flame stability*, Combustion Sci. Tech., 7 (1973), pp. 241–246. This inaptly titled paper describes acoustic waves propagating through a time invariant, spatially changing atmosphere whose properties are those of a large Damköhler number diffusion flame; a potential function satisfies an equation like (8.34) in three space dimensions, but with Ω and $\bar{a}_{fob}$ functions of spatial position rather than time.

[6] ———, *Behaviour at acoustic wave fronts in a laminar diffusion flame*, Quart. J. Mech. Appl. Math., 27 (1974), pp. 161–173. Continuation of the previous work with solutions sought for jumps in gradients on frozen acoustic wavelets in the spatially varying background.

[7] ———, *Small amplitude disturbances in an exploding atmosphere*, J. Fluid Mech., 89 (1978), pp. 343–356. Refers particularly to work in §7.3.

[8] ———, *On the evolution of compression pulses in an exploding atmosphere; initial behaviour*, J. Fluid Mech., 94 (1979), pp. 195–208. Refers especially to §9.4(iv) and discusses the shock fitting problem for initial-value or short-duration compression pulses.

[9] ———, *On the propagation of gasdynamic disturbances in an explosive atmosphere*, Prog. Aeronautics and Astronautics, 76 (1981), pp. 383–402. Extends approximate, transport-free, analysis (cf. §§7, 9) to cylindrical and spherical geometries and describes a number of possible disturbance regimes and their governing equations.

[10] ———, *On changes in the structure of plane flames as their speed increases*, Combustion and Flame, 50 (1983), pp. 125–138. Describes how transport effects begin to play a diminishing role as the speed of flow through a burner-anchored flame increases; transition to pure convection-reaction structure is completed well before compressibility effects intervene.

[11] ———, *Combustion in plane steady compressible flow: General considerations and gas-dynamical adjustment regions*, J. Fluid Mech., 136 (1983), pp. 139–166. Analyses chemical-energy release into flows in which compressibility is important; shows that convection, reaction and compressibility are the dominant mechanisms except for the interior of shock-wave-like layers in which reaction is frozen (cf. §7.3(ii)) and viscous effects balance convection; these layers are "thin" (cf. §§7.3(iii) and (iv)).

[12] J. F. CLARKE AND R. S. CANT, *Non-steady gas dynamic effects in the induction domain behind a strong shock wave*, Progress in Astronautics and Aeronautics 95, Dynamics of Shock Waves, Explosions and Detonations, American Institute of Aeronautics and Astronautics, New York, 1984, pp. 142–163. Describes the work of §10 in more detail.

[13] J. F. CLARKE, D. R. KASSOY, AND N. RILEY, *Shocks generated in a confined gas due to rapid heat addition at the boundary. I, weak shock waves; II, strong shock waves*, Proc. Roy. Soc. Lond., A, 393 (1984), pp. 309–329; 331–351. Describes wave generation in an inert gas by heat addition through a boundary (cf. §1); shows that shock-*strength* depends upon power input and sustainment upon energy input; quantifies power/strength relationship.

[14] J. F. CLARKE AND M. MCCHESNEY, *Dynamics of Relaxing Gases*, Butterworths, London, 1976. Contains information about basic equations, speeds of sound and characteristics of equations describing the flows of reacting gases, together with descriptions of the behaviour of waves propagating through equilibrium atmospheres.

[15] D. R. KASSOY, *Perturbation methods for mathematical models of explosion phenomena*, Quart. J. Math. Appl. Mech., 28 (1975), pp. 63–74. Carries the work of §9.1 through the whole event in the ambient atmosphere.

[16] D. R. KASSOY AND J. POLAND, *The thermal explosion confined by a constant temperature boundary*, SIAM J. Appl. Math., 39 (1980), pp. 412–429. Solid-body explosion, but results akin to those at local explosion in §9 above are found at ignition.

[17] J. H. LEE, *The mechanism of transition from deflagration to detonation in vapour cloud explosions*, Prog. Energy and Combustion Sci., 6 (1980), pp. 359–389. An extensive review of a variety of flame-acceleration mechanisms especially, in the present context, Shock Wave Amplification by Coherent Energy Release, or SWACER, in which energy-release by the combustion reaction takes place in synchronism with the passage of a shock wave. Correspondence of this mechanism with the work of §9.4(i)–(iv) is appealing; cf. also the work of §10; although the present analysis is only for weak waves its extensions to larger amplitudes is strongly motivated by such comparisons.

[18] G. H. MARKSTEIN, *Nonsteady Flame Propagation*. The book is made up from contributions by G. H. Markstein, H. Guénoche and A. A. Putnam. In light of developments in the past ten years (described in *Theory of Laminar Flames* by J. D. Buckmaster and G. S. S. Ludford, Cambridge Univ. Press, Cambridge, 1982) much of the theory described in the book is dated, and has been put on a firm footing and vastly extended, but the book abounds in observations and insights. It is as important a text now as it was when it was first published.

[19] T. Y. TOONG, *Chemical effects on sound propagation*, Combustion and Flame, 18 (1972), p. 207. Exploits the idea of a fixed wavelength of sound and examines the effect of some rather simple reactions, with algebraic dependence of rates on temperature for example. Finds a potential equation like (8.34) with constant coefficients.

INDEX

Acoustic disturbance, 17
  linear chemistry, 205
Acoustic instability, 33, 123
Acoustic signal, 176
Acoustic time scale, 38
Acoustic wave, 18, 142, 143, 184, 205, 208, 209
Activation energy, 7, 12, 14, 15, 16, 22, 23, 26, 106, 185, 188, 202, 239
  asymptotics, 22, 23, 35, 119, 124
  infinite, 23
  limit, nonlinear chemistry, 224
  nondimensional, 12, 14
  number, 192, 194, 195
  nonlinear chemistry, 225
Adiabatic flame speed, 19, 29
Adiabatic flame temperature, 11, 16, 19, 23, 26, 28
Adiabatic treatment for detonation, 144
Adjoint technique, 77
Age theory, 108, 120
AIM method, 57
Ambient atmosphere, 184, 185, 187, 201, 204, 219
  and chemical equilibrium, 191
  disturbance behavior, 192, 220
  equations for, 191-194
  gaseous combustion, 191-193
  nonlinear chemistry, 225
  premixed, 191
  spatial uniformity, 191
Ambient reaction rate, 203
Amplification factor, linear chemistry, 210, 218, 219
Amplitude factor, linear chemistry, 204, 218, 219
Analog, detonation, 134-145
  applications and limitations, 152, 153
Arrhenius factor, 7, 24, 61, 106, 151
  exponential, 23
  temperature, 164, 176
  reaction rate, 136, 161, 188
  kinetics, 13, 43,
Asymptotic analysis, 23-25, 28, 35, 36, 73, 220
  activation energy, *see* activation energy
  turbulence, 109
Atomic scale, 47, 51, 89
Autocatalysis, 68
Averaging, turbulence, 104

Backward acoustic characteristic, 147, 148
Beta function distribution, 116

Bifurcation points, 54
Blowoff, 118, 119
Bluff-body flame holder, 121
Boltzmann equation, 102
Boundary condition, 34, 35, 53, 60, 157, 165, 166, 167, 174, 176, 228
  deterministic, 99
  rear, 140
  sensitivity, 61
  stochastic, 99
  turbulent flow, 101
Boundary pressure, 176, 177, 178, 179
Boundary value problem, 184, 191
Boundary velocity, 140
Boundary layer formulation, 40
Bromwich contour, 206
Bulk phenomenon, 88, 89
Buoyancy, 40-42
Buoyant instability, 122
Burgers equation, 133, 152, 185, 186, 204, 205, 211, 225, 226
Burke-Schumann limit, 37
Burning velocity, 121, *see also* turbulent flame speed

Cell structure, 19, 27, 142, 161, 183
Cellular instability, 27, 28, 42
Chapman-Jouget detonation, 138, 140, 155, 156, 160, *see also* unsupported detonation
Chapman-Jouget point, 140, 142
Chapman-Jouget velocity, 140
Characteristic growth constant, 73
Characteristic length, 34, 41
Characteristic mass flux, 34
Characteristic parameter, 197
  gaseous combustion, 240
Characteristic speed, 42, 140, 153, 166, 167
Characteristic temperature, 34
Charles' law, 41
Chemical activity, 186
Chemical composition field, 184
Chemical concentration, 55
Chemical energy, gaseous system, 183
Chemical equilibrium, turbulent combustion, 111
Chemical instability, 67, 73
Chemical kinetics, 7-9, 55, 89, 98, 102, 107
Chemical reactor, 71, 121
Chemical species, 65, 71, 72, 99
  concentration, 64, 67, 73
  conservation, 101, 188

247

Chemically frozen flow, 198
Chuffing, 33
CJ, see Chapman-Jouget
Classical wave equation, 223
Closure, 101, 105, 128
  turbulence, 101, 102
Coherence, 187
Cold boundary difficulty, 11, 44, 191
Cole-Hopf reduction, 205
Collision cross section, 88
Colored noise, 67
Combustible gases, finite amplitude waves in, 183–245
Combustion energy number, 194
Combustion equations, 50, 59
Combustion field, 10, 38
Combustion theory, fundamentals of, 3–46
Composition, 173, 175
Compression wave, 229, 230
  gas dynamical, 187
  linear chemistry, 210, 211
Concentration, 13, 16, 67, 68, 76
  as dependent variable, 77
  mean, 68
Concentration correlation function, 68
Concentration gradient, 5, 55
Concentration profile, 59, 74, 75, 89
Concentration trajectory, 68
Conduction coefficient, 27
Conduction-diffusion scale, 29
Connectivity of reactions, 89
Conservation condition, 135
Conservation equation, 6, 99, 124, 125, 188
  small disturbance, 236
  nonlinear, 104
Conserved scalar, 128
  turbulence, 107
Constant density model, 26, 30, 32, 101, 106, 174
Constant diffusivity, 101
Control, 49, 64, 71, 90, 91
Cordite, 35
Correlation, 84
  diffusion and rate, 81
Coulombic interaction, 88
Countergradient diffusion, 107, 116
Coupling, 54, 89, 185
  and network theory, 89
  chemical, 89, 185, 186, 189
  chemical radicals, 71, 72
  chemistry and wave processes, 185, 189
  diffusion and kinetic, 80
  gas dynamical, 186, 187
Covariance, 66, 84, 85
Curvature, 19, 20, 21, 124, 143
  laminar flame, 121
  reaction sheet, 124

Dämkohler number, 36, 110, 112, 113, 124
Darrieus-Landau analysis, 19, 33

DDT, see deflagration to detonation transition
Dead water hypothesis, 22
Deflagration, 16, 17, 25, 26
Deflagration initiation, detonation, 144
Deflagration to detonation transition, 44, 144, 145, 187
Deflagration wave, 10–12, 19, 23, 25, 26, 33, 133, see also plane flame
Delta function, 25–35, 115, 124
  Dirac, 25, 76, 238
  Kronecker, 57
Derived sensitivity, 76–82
Deterministic system, 52, 53, 69
Detonation, 9, 17, 44, 133, see also deflagration
  current theory, 134, 142–145
  galloping, 161, 162, 164
  in a tube, 143, 157
  in miniature, 133–181
  initiation, 144, 172–180
  physical system for, 134–145
  spherical, 142
  steady, 154
  two-dimensional, 144
Detonation front, 138, 143
Detonation Hugoniot, 142
Detonation pressure, 140
Detonation velocity, 140, 143
Detonation wave, 35, 133, 152, 183
  ZND model, 138
Diffusion approximation, 105
Diffusion coefficient, 27, 53, 82
Diffusion dominated process, 185
Diffusion flame, 7, 16, 35–42, 111, 113, 117, 120
  chemical equilibrium, 118
  distributed reaction, 119
  instability, 122
  reaction sheet, 115–119, 127
  turbulent jet, 120
Diffusion process, 80
Diffusive mass flux, 188
Diffusive term, 152
Diffusive-thermal instability, 123
Diffusivity, 209
Dimensionless equations, gaseous combustion, 193–195
Dirichlet problem, 61
Dispersive wave system, 186, 223, 242
Dissipation length, 116
Distinguished limit, 24, 200, 201, 220
Distributed parameters, 49, 50
Distributed reaction regime, 127
Disturbance amplitude, 185
Dynamic equilibrium, 57, 58
Dynamical equation, 50, 54

Eddy breakup model, 105, 121
Eigenvalue detonation, 141, 156
  unsupported, 156
Eigenvalue-eigenvector analysis, 91

INDEX

Elementary sensitivity, *see* sensitivity
Elliptic problem, 29
Energy balance equation, 53, 61
Energy conservation, 61, 136, 188
Energy equation, 4, 11, 13
Energy flux, 4, 6, 61, 188
Enthalpy, 4, 5, 107, 111, 188, 189
Entropy, 17, 136
  analog, 151
Equal-areas formula, 213
Equation of motion, 53, 166
Equation of state, 4, 10, 11, 38, 41, 137, 145, 161, 165, 166, 176, 189
  analog, 149, 150, 152
  polytropic gas, 177
  thermal, 136
  turbulent combustion, 111
Equilibrium constant, 8
  analog, 151
Equilibrium dependence, 117
Equilibrium limit, 37
Equilibrium sound speed, 153, 154
Error analysis, 48, 82, 85
  sensitivity analysis, 82
Error propagation, 51
ESCIMO theory, 108
Euler equation, 19, 103
  inviscid compressible flow, 134, 135
  nonreactive, 145
Exothermic reaction, 15, 184
  gaseous combustion, 196
Expectation, 66, 67
Extinction, 31, 38, 45, 118
  fuel drop, 45
  laminar flame, 127
  local, 118
Extracted parameter, 82

Far field, 28, 123, 126, 175
Favre average, 106
Feature parameter, 75, 76
Feature sensitivity analysis, 52, 73-77, 86
Fick's law, 5, 188
Finite amplitude waves in combustible gases, 183-245
Finite difference method, 88
  turbulence, 103
Finite element method, turbulence, 103
Finite excursion, 91, 92
Fitting, 74, 75
Fizz burning, 33-35
Flame bubble, 28
Flame curvature, *see* curvature
Flame length, 18, 19, 117
Flame profile, 82
Flame propagation, 28, 44
Flame sheet, 23, 24, 26, 29, 30, 37, 39, 40, 122, 134
  spherical, 28
  thickness, 39
Flame sheet model, 32, 37, 117
Flame speed, 11, 12, 22, 26, 31, 82
  negative, 31, 32
Flame stretch, 123, 124
  reaction-sheet regime, 123
  weak, 124
Flame structure, 17, 18, 23
Flame tip, 22, 29
Flicker, 40
Flow
  statistical properties, 101
  turbulent variable density, 106
Flow characteristic, 174
Flow field, 30, 183
Fluctuation dissipation theorem, 66, 67
Fluid flow, equations of, 98
Fluid instabilities, 99
Fluid mechanics, 26, 32, 38, 42, 47, 121
  turbulence, 107
  velocity, upstream, 124
Fluid dynamic instability, 123
Flux, *see* specific parameters
Fokker-Planck equation, 68
Following flow, 160
Forced convection, 40
Forward acoustic characteristic, 147, 148
Fréchet derivative, 63
Frequency dependence, 59
Fresh gas, 19, 21
Front, 19
Frozen chemistry, 210, 211
Frozen Rankine-Hugoniot shock, 212
Frozen sound speed, 136, 143, 150, 153, 190, 222, 223, 237
  adiabatic, 223
  nonlinear chemistry, 224
Frozen specific heat, 189
Fuel drop, 35
Fuel-lean (rich) mixture, 12, 13, 27, 28
Functional derivative, 63, 64, 65, 70
Functional equation, turbulent combustion, 99
Functional sensitivity analysis, 49, 62-65
Functional variation, 63, 70
Fundamental equations of combustion theory, 3-7

Gallop, 161, 162, 164
Gas dynamics, 183-191
Gas phase, 33, 34
Gas velocity, 188
Gaseous combustion
  basic equations, 188-190
  transient phenomena, 187
Gateux derivative, 63
Gauss law, 10
Gaussian stochastic process, 66
Geometrical acoustics, 142, 143
Geometrical detonation dynamics, 143, 144

Global mapping, 48, 55, 74, 87, 91, 92
  and design and control, 91
Global momentum balance, 21
Gradient method, 54, 68
Gradient sensitivity analysis, 91
Granularity, 142, 145, 164
Green's function, 57, 62, 64, 69–71, 76
  adjoint, 77
  reduced, 77
  system, 57, 64, 68
  matrix, 70, 72, 73, 91
Group generator, 92
Group velocity, 153

Hamiltonian, molecular, 88
Harmonic disturbance, 208
Harmonic wave, 184, 185
Heat conduction, 33, 190
Heat flux, 26
Heat of formation, 5, 6
Heat of reaction, 150, 157, 172, 175
Heat, specific, *see* specific heat
Hierarchical modelling, 88, 89
High Mach number, 33
High temperature reaction, 16
High intensity small-scale regime, 121
Homogeneous mixture, 35, *see also* premixed flame
Hopf bifurcation, 27
Hugoniot curve, 139
Hydrodynamic front, 20, 21
Hydrodynamic model, 19, 27
Hydrodynamic problem, 20, 22
Hydrodynamic stability, 19, 40, 122, 123, 134, 161
Hydrogen flame bubble, 28, 32
Hydrogen-oxygen flame, 82
Hyperbolic problem, 175

Ideal gas, 136
Ignition, 16, 38, 39, 118, 186
  ambient atmosphere, 186
  hydrogen, 14
Ignition line, 118
Ignition point, 38
Ignition problem, 15
Ignition source, 191
Ignition temperature, 44
Ignition time, 192
Induced velocity, 39, 42
Induction time, 192, 195, 214, 219
Induction zone, 44
Infinitesimal excursion, 92
Infinitesimal perturbation, 19, 35
Inhomogeneity, 14, 58, 64
Initial condition, 38, 53, 56, 62, 67, 76, 147, 157, 167, 168
  deterministic, 99
  stochastic, 99
Initial disturbance, 21

Initial value problem, 10, 184
  turbulence, 103
Initial velocity, 38
Initiation of detonation, 142
Initiation transient, 175
Input parameter, 50, 51, 74, 75, 87, 164, 167
Intermolecular potential, 89
Internal energy, 135
Intrinsic instability, 35
Invariant observation, 51
Inverse, generalized, 85
Inverse kinetics, 82
Inverse sensitivity coefficient, 79
Inversion, 74, 82, 89, 161
Inviscid equation, 103
Isentrope, 138
Isothermal system, 15, 101
Iterative algorithm, 60

Jacobian, 56, 58, 87
Jet, 118, 119
Joint probability density function, 111, 117
Jump, 150, 154

Kinematic argument, 19
Kinematic incompatibility, 22
Kinematic viscosity, 110
Kinetic energy, 9, 133
Kinetic equations, deterministic, 67
Kinetic measurement, 82
Kinetic model, 8, 12–17, 32, 43, 44
Kinetic process, 12, 13, 51, 59, 74, 80, 89
Kolmogorov, 112, 113
Korteweg-de Vries equation, 133

Lagrangian coordinate, 145
Laminar flame, 6, 97, 113, 122, 123, 124
  curvature, 121
  diffusion, 118, 119
  flow, 109, 112
  propagation, 113
  reaction sheet regime, 121
  thickness, 125
  wrinkled, 121, 122, 123
Langevin equation, 66
Laplace transform, 167, 168, 170, 206
Large activation energy, 43
Large disturbance, gaseous combustion, 238–242
  strong shock wave, 238
Large scale turbulence, 113
Latent heat, 36
Latent-image rate, 169
Least squares, 75, 82, 83, 85
Lewis number, 12, 23, 26, 28, 31, 119, 125, 126
  infinite, 28
Lie algebra, 91
Lie technique, and global mapping, 91, 92, 93
Liftoff, 118, 119, 120
Limit cycle, 90

Limiting regime, turbulent combustion, 113
Linear chemistry, gaseous combustion, 203–217
  acoustical disturbances, 205
  amplification factor, 210
  change in composition, 215, 216
  compression wave, 210, 211
  frozen chemistry, 210
  general results, 204
  nonlinear disturbance, 209
  weak nonlinear wave, 213, 214
  weak shock, 211–216
Liouville equation, 102
Long time scale, gaseous combustion, 191
Longitudinal instability, analog, 162
Low frequency oscillation, 33
  and sensitivity analysis, 90
Lumping, 52, 90, 91
  and sensitivity analysis, 52, 90
Lyapunov coefficient, asymptotic, 73

Mach number, 9, 44, *see also* small Mach number approximation
  zero, 126
Mass action, law of, 8
Mass conservation, 11, 36, 106, 136, 188
Mass diffusion coefficient, 188
Mass equation, 145, 146
Mass flux, 26, 29, 34, 122
  diffusive, 188
Mass fraction, 4, 6, 9, 12, 36, 102, 110, 135, 188, 191
  gradient, 102
Material derivative, 135
Mathematical modelling, perspective, 47–52
Maxwellian distribution, 7
Mean residence time, 108
Mean square fluctuation, 115
Mechanical model for analog, 148, 149
Mesh point, 52, 56, 88
Method of steepest descents, 206, 207
Mixing, 107, 115, 119
Mixture fraction, 111, 112, 117
Model equations, 53
  missing parameter, 86
Model performance, 51
Model uncertainty, 85
Modecular diffusion, 102
Moment method, 107, 115, 116, 119, 128
  turbulence, 104–107
Momentum, 53, 133
Momentum balance, 21
Momentum conservation, 136, 188
Momentum equation, 9, 10, 39, 145, 146, 201
Monte Carlo method and global mapping, 91, 93
Multiple-scale method, 126

Navier-Stokes equation, 98, 107
  compressible, 134
Navier-Stokes turbulence, 99

Networks and sensitivity analysis, 89, 90
Neumann problem, 61
Newton algorithm, 60
Noise, 67
Nonacoustic oscillation, 33
Nonconstant coefficient, 26
Nondimensional system, 11, 18, 34
Nondimensional turbulence intensity, 112
Nonequilibrium chemical reaction, 119
Nonlinear chemistry, gaseous combustion, 224–238
  ambient atmosphere, 224, 225
  disturbance, 228, 235–238
  frozen sound speed, 224
  general solutions, 226
  local, 227–231
  perturbation equations, 225, 235
  shocks, 227, 233–235
  short pulse, 227–232
  temperature, 236, 237
Nonpremixed system, 112, 119
  turbulent combustion, 110, 111
Nonreacting flow, 109, 115
Nonreacting turbulence, 102, 108, 109
Nonreactive analog, 148, 149
Numerical computation, 17, 20, 52, 54, 56, 87, 88, 241
  dynamic code, 144
  integration, 65
  sensitivity coefficients, 87
  turbulence, 103–104, 128
Nyquist diagram, 170

Objective function sensitivity analysis, 52
Observable, 49, 50, 66, 73
One-dimensional equations, 17, 34
One-dimensional flow
  detonation, 134, 142
  unsteady, 188
One-dimensional perturbation, 29, 161
One-reaction system, detonation, 141
One-step kinetics, 11, 23, 25
Optimizing function, 82
Oscillating system, 58, 59
Output, 50, 51, 167
  as function of species, 73
  physical, 74
Output function, 164
Output parameter, 74, 86
Overdriven detonation, 140, 155, 160, 162

Padé approximant, 92
Parameter covariance matrix, 84
Parameter space, 51, 54, 91, 93
Parameter values, nominal, 63
Parametric response surface, 65
Parametric sensitivity analysis, 52
Particle velocity, 135, 140
Performance index, 50

Periodic reaction, 58, 59
Perturbation equation, 26
Perturbation method, 124
  turbulence, 109
Perturbation theory, 175
Physical process, model for, 74
Physical stability, 71, 73
Physical variable, 42
Piston problem, 157, 177, 191, 196, 214, 238
  compared to reaction/wave interaction, 183, 184
  detonation, 134
  gaseous combustion, 195
Plane disturbance, 26
Plane flame, 10-11, 26, 28, 29, 122, *see also* deflagration wave
Plane shock 172
Plane wave, 186
Plug-flow reactor, 108
Pollutant, 117
Polytropic gas, 176
Prandtl number, 125, 194, 195, 208
Prandtl theory, mixing length, 105
Premixed atmosphere, 191
Premixed flame, 10, 22, 23, 61, 111, 123
  distributed reaction, 120, 121
  one-dimensional, 10, 82
  reaction sheet, 121-127
Premixed system, 112, 120
  turbulent combustion, 110, 111
Pressure, 18, 135, 175
  ambient, 192
Pressure gradient, 10, 39
Pressure perturbation, 175, 178
Pressure wave, 183
Probability density function, 99-103, 115, 128
  approximation of, 104, 120
  evolution of, 99-103
  joint, 102
Probability distribution function, 66, 68
  concentration, 65, 66
Propagation time, finite, 65
Propagation velocity, eigenvalue detonation, 156
Propellant, 33
Pulsating flame, 58, *see also* travelling wave
Pulsating front, 28
Pulsating instability, 35
Pyrolysis, 33-35, 86

Quantum dynamics, 88
Quantum mechanics, 47
Quenching, 39, 45, *see also* extinction

Random variable, turbulent combustion, 99
Random-vortex method, 103
Rank reduction, 91
Rankine-Hugoniot discontinuity, 186
  frozen shock, 212
Rankine-Hugoniot relation, 136, 138
Rarefaction wave, 154, 156, 160

detonation theory, 138
following, 138, 140
Rate constant, 56, 61, 64, 80, 84, 88
  analog, 151
  statistical values, 67
Rate equation, 160, 165, 169, 173
  analog, 150
  detonation, 139
  physical, 148
Rayleigh line, 137, 139, 140, 147, 155, 156, 160
RCL network, 163
Reacting gas mixtures, 99
Reaction constant, forward (reverse), 58
Reaction front, 183
Reaction rate, 7, 98, 164, 165
  backward (forward), 67
  detonation, 139
  turbulence, 102
Reaction sheet, 113, 115, 116
  finite thickness, 127
  limit, 116
  multiple sheets, 113
Reaction sheet regime, 113, 115, 121, 126, 127
  instability, 122, 123
  flame stretch, 123
Reaction surface, connected, 113
Reaction time, 110
Reaction zone, 23, 25, 29, 38, 39, 138, 140, 154, 170, 172, 173
  detonation, 143
  detonation theory, 138
  steady, 159
  profile, 154
Reaction, exothermic, *see* exothermic reaction
Reaction-based amplification, 208
Reaction-diffusion, 53, 62, 65
Reaction-diffusion model, 53, 71, 80
  steady state, 53
  one-dimensional, 72
  temporal case, 53
Reaction/wave interaction, 183
Reactive balance, 16
Reactive flow, 187
Reduced model, 85, 86
Reduced system, 10
Reference composition, 173, 174, 175
Reference flow, 173
Reflected wave, 176
Reflection coefficient, 142, 177, 178
Regime
  identification of, 113
  limiting
    assessment of, 115-127
    diffusion flame with reaction sheet, 115-119
    turbulent combustion, 109-115
Residence time, 108
Resonant matching, 58
Reynolds number, 42, 185, 194, 195, 199, 200, 201, 208

INDEX

Taylor scale, 112
turbulence, 110
Reynolds stress, 105
Right stability boundary, 33–35, *see also* stability boundary
Rocket motor, 17, 33–35
Runaway, 14, 38, 73
Runge-Kutta integration, 170

Scalar dissipation, 120
Scalar field, 107
Scaling
  detonation, 157
  gaseous combustion, 198
  nonlinear, 75
Schmidt number, 194, 195
Scrambler, 49
Second moment, 105
Second order sensitivity coefficient, 55
Secularity, 59
Semi-characteristic coordinate, 185, 197
Sensitivity analysis, 17, 44, 47–93
  and error estimation, 82–85, 82
  and experimental data, 82–85, 82
  and hierarchical modelling, 88, 89
  and lumped system reduction, 90
  and network theory, 89, 90
  applications, 51
  computational aspects, 87–88, 87
  constant parameter, 63, 64
  derived, 76–82, 77
  elementary, 62, 63, 69, 71, 80
  feature, 52, 73–77, 85
  first order, 78
  functional, 62
  gradient, 91
  missing model components, 85–86, 90
  objective function, 73–77
  role in mathematical modelling, 47
  techniques, 48, 53–83
  temporal systems, 66
Sensitivity coefficient, 51–91
  backward (forward), 57
  concentration profile, 75
  derived, 79, 80
  elementary, 54–62, 69, 78, 79, 83, 90
  feature, 76
  first order, 55, 78, 83. 85
  frequency, 59
  mesh, 88
  nonlocal, 88
  second order, 83
Sensitivity density, 64, 65
Sensitivity equation, 54, 87
  inhomogeneous, 60
  linear, 60
Sensitivity matrix, 78, 79, 84
Shock, 136, 138, 144, 153, 157, 160, 165, 168, 169, 177

constant velocity, 154
plane, 172
initiation, detonation, 144
initiation problem, 145, 172
Shock density, 136, 165, 167, 169
Shock frame, 139, 158, 159, 165, 174, 176
Shock Hugoniot, 137, 138, 147
Shock jump, 150, 154, 156
Shock path, 168, 215
Shock pressure, 161
Shock relation, 165, 166
Shock state, 136
Shock strength, 164, 215
  nonlinear chemistry, 233
Shock velocity, 136, 138, 139, 158, 165, 173, 175
Shock wave, 134, 183, 187
  initiating, 144
  linear chemistry, 213
  nonlinear chemistry, 227, 229, 230, 231
Short pulse approximation, 221
  nonlinear chemistry, 227
Short time scale, gaseous combustion, 191
Shrödinger equation, 88
Signal velocity, 183
Sivashinsky model, 21
Small Mach number approximation, 9–10, 17–19, 38
Small perturbation, 12, 201, 202, *see also* perturbation
  uni-directional waves, 199–203
Small wave number expansion, 27
Small scale turbulence, 113
Sound speed, 18, 147, 174
  frozen, 190
  local, 186
Space-time problem, 53, 62, 70
  sensitivity analysis, 62
Spatial homogeneity, turbulent combustion, 101
Specific heat, 3, 5, 6, 136
  frozen, 189
Spectral method, 103
Stability, 12, 19, 26, 48, 69–71,
  detonation, 142, 161–171
Stability boundary, 27, 28
Stable limit cycle, 58
Stagnation point, 31
Stand-off distance, 31
State curve, 137
State dependence, 165
State variable, turbulent combustion, 99
Stationary flow, 104, 116
Stationary front, 19
Statistical error, 84
Steady detonation, 138, 156, 159
  analog solution, 156
  stability, 157
Steady state solution, 54, 59, 90
  perturbations, 142
Steady state velocity, 38, 39

Stewart model, 33
Stiff system, 54
Stirred reactor regime, 113, 118, 119, 121, 127
Stochastic equation, 52, 66, 67, 68
  turbulent combustion, 99
Stochastic system, 65-69
Stoichiometrics, 8, 9, 12, 117
Strain, 30-32, 124
Strange attractor, 99
Stratonovich, 67
Strong point, 140
Strong shock, 147
Strong shock wave, gaseous combustion, 238-242
Strong stretch, 127
Supported detonation, 140
System Jacobian, 60
System parameter, 50, 51, 59, 66, 86
System sensitivity, 50, 54, 61
  parametric, 73
System stability, 51, 64

Taylor scale, 112
Taylor series, 55, 84, 91
Temperature, 14, 18, 28, 38, 39
  adiabatic flame, 11
  dependence, 7, 15, 77
  sheet, 41
  perturbation, 240, 241
Temperature gradient, 24, 25, 187
Temperature profile, 50, 74, 89
Temporal system, 56, 58, 60, 65, 70, 71, 92
Thermal factor, analog, 151
Thermal flux, 10
Thermal ignition, 14, 15, *see also* ignition
Thermal model, 15, 16
Thermicity, 150
Thermite, 28, 33
Thermodynamics, 74, 78, 82, 115, 185
  analog, 151
  and chemical equations, 185
Time history, 74
Tracer experiment, 108
Trajectory, deterministic, 69
Trajectory sampling, 67
Transfer function, 164, 167, 169
Transport, 47, 51, 74
  gaseous combustion, 196, 201
  turbulent, 107
Transport coefficient, 41, 82
  molecular, 110
Travelling wave, 27, 175, 177, 183
Triple shock, 142
Tube problem, 19, 20, 21
Turbulence, 32, 44
  as stochastic phenomenon, 98
  large scale (small scale), 121, 122
Turbulence intensity, 113

Turbulence modelling, 7, 101, 105
Turbulent combustion, 96-131
  analytic approaches, 98-99
  definition, 98
  multifaceted nature, 97
  prospects, 127, 128
  stirred reactor regime, 121
Turbulent diffusion, 40, 105, 107, 122
Turbulent flame, stationary, 125
Turbulent flame speed, 121, 122
Turbulent flow, 99, 122
  confined, 121
  high-intensity, 121
  premixed, 121
  variable density, 106
Turing system, 26
Two-dimensional detonation, 143, 144

Uncertainty estimate, 85
Undisturbed solution, 26
Undisturbed variables, 18
Uni-directional behavior, 232
Uni-directional propagation, 186
  breakdown, 217-224
  gaseous combustion, 217
  nonlinear chemistry, 225
  small perturbation, 202
Uni-directional wave, 185, 203
  gaseous combustion, 195-199
  small perturbation, 202
Unstable front, 28
Unstable wave field, 186
Unsupported detonation, 138, 140, 160, 162, 170

Variable density, 106, 128
Velocity field, 26, 28, 50, 101
Viscosity, 9, 126, 152, 190, 201, 205, 208
  coefficient, 188
Von Neumann point, 140
Vorticity, 20, 21, 108

Wave field
  dispersive, 186
  unstable, 186
Wave head, 153, 207
Wave number, 26, 27
Wave propagation, 44, 185
  gaseous combustion, 196
Weak detonation wave, 35
Weak point, 141
Weak shock, 211
Weak stretch, 124, 126
Weak turbulence limit, 115
White noise, 67

Zeldovich model, 15, 16, 17, 33
Zeldovich number, 125
Zero-crossing problem, 117
ZND model, 138, 154, 172, 187

Ollscoil na hÉireann, Gaillimh